Laser Applications
in Chemistry

NATO ASI Series

Advanced Science Institutes Series

A series presenting the results of activities sponsored by the NATO Science Committee, which aims at the dissemination of advanced scientific and technological knowledge, with a view to strengthening links between scientific communities.

The series is published by an international board of publishers in conjunction with the NATO Scientific Affairs Division

A	Life Sciences	Plenum Publishing Corporation
B	Physics	New York and London
C	Mathematical and Physical Sciences	D. Reidel Publishing Company Fordrecht Boston, and Lancaster
D	Behavioral and Social Sciences	Martinus Nijhoff Publishers
E	Engineering and Materials Sciences	The Hague, Boston, and Lancaster
F	Computer and Systems Sciences	Springer-Verlag
G	Ecological Sciences	Berlin, Heidelberg, New York, and Tokyo

Recent Volumes in this Series

Volume 98 —Quantum Metrology and Fundamental Physical Constants
edited by Paul H. Cutler and A. A. Lucas

Volume 99 —Techniques and Concepts in High-Energy Physics II
edited by Thomas Ferbel

Volume 100—Advances in Superconductivity
edited by B. Deaver and John Ruvalds

Volume 101—Atomic and Molecular Physics of Controlled
Thermonuclear Fusion
edited by Charles J. Joachain and Douglass E. Post

Volume 102—Magnetic Monopoles
edited by Richard A. Carrigan, Jr., and W. Peter Trower

Volume 103—Fundamental Processes in Energetic Atomic Collisions
edited by H. O. Lutz, J. S. Briggs, and H. Kleinpoppen

Volume 104—Short-Distance Phenomena in Nuclear Physics
edited by David H. Boal and Richard M. Woloshyn

Volume 105—Laser Applications in Chemistry
edited by K. L. Kompa and J. Wanner

Volume 106—Multicritical Phenomena
edited by R. Pynn and A. Skjeltorp

Series B: Physics

Laser Applications in Chemistry

Edited by

K. L. Kompa

and

J. Wanner

Max Planck Institute for Quantum Optics
Garching, Federal Republic of Germany

Plenum Press
New York and London
Published in Cooperation with NATO Scientific Affairs Division

Proceedings of the NATO Advanced Study Institute on
Laser Applications to Chemistry,
held June 27–July 11, 1982,
at San Miniato, Italy

Library of Congress Cataloging in Publication Data

NATO Advanced Study Institute on Laser Applications to Chemistry (1982: San
   Miniato, Italy)
   Laser applications in chemistry.

(NATO ASI series. Series B, Physics; v. 105)
   "Proceedings of the NATO Advanced Study Institute on Laser Applications in
Chemistry, held June 27–July 11, 1982, at San Miniato, Italy—T.p. verso.
   Includes bibliographical references and index.
   1. Lasers in chemistry—Congresses. I. Kompa, K. L. (Karl L.). II. Wanner, J. III.
North Atlantic Treaty Organization. Scientific Affairs Division. IV. Series: NATO
advanced science institutes series. Series B, Physics; v. 105.
QD701.N36   1982   540   83-27047
   ISBN-13: 978-1-4612-9697-3   e-ISBN-13: 978-1-4613-2739-4
   DOI: 10.1007/978-1-4613-2739-4

PREFACE

This volume contains lectures and seminars presented at the Nato Advanced Study Institute on "Laser Applications to Chemistry" held at San Miniato (Pisa) Italy, June 27 - July 11, 1982.

We would like to give our recognition to all who contributed to the superb scientific quality and to the stimulating atmosphere of this summer school. In particular, we thank all speakers and participants in the discussions. We acknowledge the great efforts of Tito Arecchi as the director of the school, and the assistance of Mrs. Maria Bonaria Petrone and Mrs. Giovanna Ravini in the organization of the conference. On behalf of all participants we thank Miss Iva Arecchi for the friendly care she has taken and for her profound guidance through its history and the arts of the Toscana during the excursions.

We are in particular indebted to Giacinto Scoles who organized this meeting together with Karl L. Kompa. Unfortunately, due to health reasons, Giacinto Scoles had to give up the idea of editing the proceedings of this meeting. Naturally, the change in the editorial staff caused a delay in the preparation of this volume. The subject of laser application to chemistry has not been reviewed comprehensively in recent years. Many of the lectures and seminars presented in San Miniato had the character of review articles. Therefore we feel that the material contained in this volume has not lost any of its actuality.

We are indebted to the Scientific Affairs Division of the North Atlantic Treaty Organization for the financial support of this meeting.

Munich, July 1983

K.L. Kompa
J. Wanner

CONTENTS

Introduction. 1
    K.L. Kompa and J. Wanner

LASER AND RELATED LIGHT SOURCES

Laser Sources for Chemical Experiments. 15
    K.L. Kompa

High Power Optically Pumped Mid-Infrared
        Molecular Gas Lasers. 25
    R.G. Harrison

Synchrotron Radiation and Laser Excitation Sources
        for Studies of Intramolecular Dynamics. 35
    S. Leach

Enhancing Synchrotron Radiation: Wigglers, Undulators
        and the Free Electron Laser 47
    S. Leach

LASER APPLICATIONS TO ANALYTICAL CHEMISTRY

Analytical Chemistry Methods Based on
        Absorption of Laser Light 57
    J.C. Wright

Laser Excited, Fluorescence Methods
        in Analytical Chemistry 67
    J.C. Wright

Multiphoton Ionization in Analytical Chemistry. 75
    J.C. Wright

Nonlinear Spectroscopic Techniques and their
        Applications to Analytical Chemistry. 81
    J.C. Wright

Laser Measurements of Trace Gases in
        the Atmosphere and in the Laboratory.   89
    W. Krieger and H. Walther

                  SPECTROSCOPIC AND DYNAMICAL STUDIES

VUV Laser Spectroscopy of Atomic and Molecular Hydrogen . . . 103
    K.H. Welge

Laser Spectroscopy of Molecular Ions. 117
    A. Carrington

Photodissociation Dynamics Experiments with NO_2 123
    K.H. Welge

Collision Induced Mode Selective Energy
        Transfer in Methylfluoride. 133
    R. Stender and J. Wolfrum

Infra-Red Laser Induced Energy Distributions
        in Polyatomic Molecules 141
    R.G. Harrison

Dynamics of Multiphoton Excitation of Polyatomic
        Molecules by Means of One or Two IR Laser
        Frequencies 151
    R. Fantoni, E. Borsella, and A. Giardini-Guidoni

Multiphoton Ionization and Fragmentation of
        Polyatomic Molecules. 161
    F. Rebentrost

Multiphoton Selective Excitation: on the Role of
        Chaotic Dynamics with many Basins of Attraction 171
    F.T. Arecchi

Two Applications of Lasers:
        I. Multiphoton Excitation of Chemical Reactions
        II. Mode Specific Excitation of Bimolecular
                Reactions. 183
    G.C. Pimentel

State-to-State Chemical Kinetics Studied with
        Laser-Induced Fluorescence. 193
    J. Wanner

The Effect of Vibrational and Translational
        Excitation in Atom-Molecule Reactions 199
    K. Kleinermanns and J. Wolfrum

CONTENTS

IR Laser-Induced Desorption, Reaction,
and Ionization Processes at Surfaces. 207
P. Hess

Study of Molecule-Surface Interaction by Lasers 215
W. Krieger and H. Walther

Laser Photolysis Studies of Quinone Reduction by
Chlorophyll in Homogeneous and Heterogeneous
Systems 225
F. Castelli

APPROACHES TO LASER SYNTHESIS

Laser Initiated Free Radical Chemistry. 239
D.J. Perettie, J.C. Stevens, and J.B. Clark

Laser Induced Selective Decomposition Reactions 245
D.J. Perettie, S.M. Khan, J.B. Clark, and
J.M. Grzybowski

Photoinitiated Catalysis by Transition Metal Carbonyls. . . . 251
D.J. Perettie, M.S. Paquette, R.L. Yates, and
H.D. Gafney

Formation of C_2H_3Cl by Laser-Induced Radical Chain
Reactions 259
M. Schneider and J. Wolfrum

List of Participants. 267

Index 271

INTRODUCTION

K.L. Kompa and J. Wanner

Max-Planck-Institut für Quantenoptik
8046 Garching, Germany

Laser chemistry in its historical development can be considered as an off-spring of laser spectroscopy. Consequently, the first concepts for this type of chemistry were formulated by physicists. These envisaged the use of single quantum excitation in reagents to lead to very specific products in a very controlled way (comp. Fig. 1). This type of single-level chemistry has developed into an important aspect of laser chemistry, although it has become more fruitful for diagnostic and analytic than for synthetic chemistry.

The potential of the laser as a tool to carry out chemical analysis up to a high level of sophistication has been exploited widely. All chemical compounds except the rare gases result from one or the other type of atomic or molecular interaction. Thus the broad diagnostic potential of the laser can be visualized using a simple chemical reaction as an example. In Fig. 2 it is shown how the laser can be used to probe different stages of the reaction process.

Traditionally, chemical analysis is concerned with the identification and concentration measurement of the final product of a reaction. Soon after the advent of the laser it was realized that it can advantageously replace light sources in conventional absorption and fluorescence experiments. Furthermore the intrinsic properties of the laser led to the development of new types of analytical techniques such as optoacustic, intracavity, and thermal lensing spectroscopy, multiphoton ionization and various methods based on nonlinear spectroscopy. The spatial collimation of the

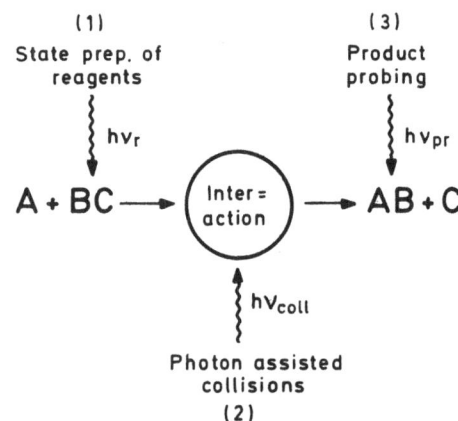

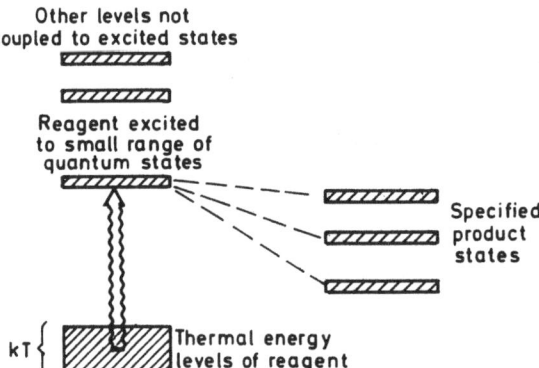

Fig. 1:   Principle scheme of state specific chemistry. The upper
picture shows the three basic types of laser interaction
with a reactive system including the influence a laser
field can have on the reactive collision complex itself
(photon assisted reactive collision). As the lower picture indicates this kind of laser controlled reaction
can only be considered if suitable excited states exist
which do not couple extensively to neighboring levels.

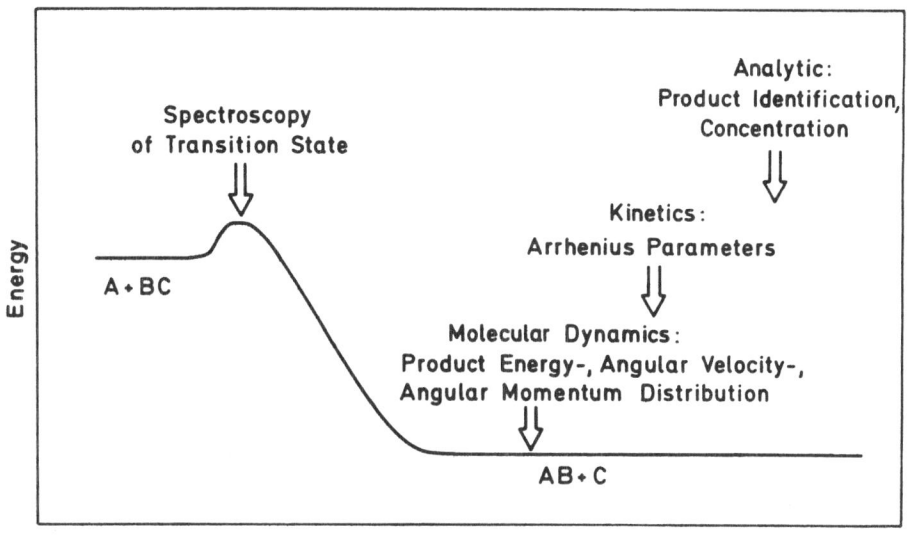

Fig. 2:   Probing of a simple chemical reaction exemplifies the
          potential of the laser as a diagnostic tool. In this pic-
          ture the reaction coordinate stands for the time scale
          of the quilibration of the reaction products. The product
          can be analyzed i) state specifically in statu nascendi
          ii) in its macroscopic rate of formation iii) in its
          final constitution and stationary concentration.
          Recently insight into the transition state has been
          gained by laser excitation methods.

laser radiation in addition allowed the transfer of some of these
techniques from the laboratory into the atmosphere in order to study
atmospheric properties and pollution. In several experiments it
has been demonstrated, that these methods can be applied also for
monitoring of immisions from chemical plants.

   Laser diagnostics in classical kinetics is concerned with the
evaluation of product concentrations as a function of time. Thereby many of the techniques listed above have been applied. The
measurements yield the Arrhenius parameters. The Arrhenius law up
to now has proven to be a valid concept for the description of ma-

croscopic rate processes though it fails to give insight into the
detailed dynamics of the collision event. This information can be
obtained if the reaction products are probed at an earlier stage
of the reaction coordinate. Due to the observation immediatly
after formation they still exhibit their internal nonequilibrium
distribution.   Though   basically   several   of   the   analytical
techniques as outlined above can be applied to experiments in the
regime of Molecular Dynamics the technique of tunable laser induced fluorescence product state analysis has become of out-
standing importance. The quantities which can be measured are the
internal product state distribution as well as the angular
velocity distribution. Both contain the necessary information for
the construction of the potential energy surface.

    The new frontier in this field is concerned with the
spectroscopy of the transition state. Yet being in its infancy

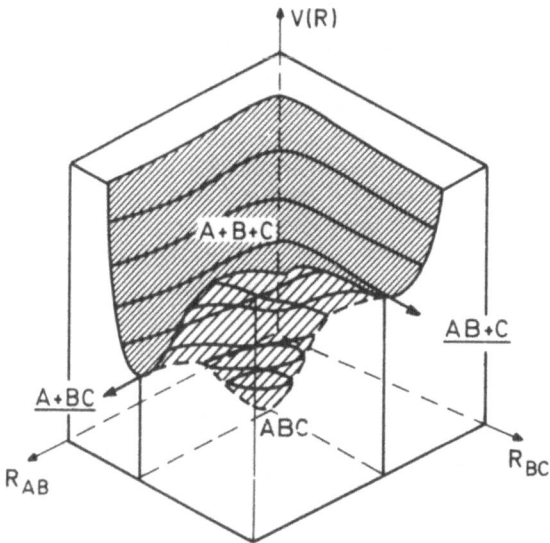

Fig 3:   Idealized picture of vibrational photochemistry represen-
          ting the chemist's dream of breaking a strong bond in a
          molecule while leaving a weak one intact. The nuclear
          coordinates corres ponding to the AB (R_{AB}) and BC (R_{BC})
          bonds are treated independently. Thus vibrational exci-
          tation along either of these coordinates would lead
          eventually to breaking of either one of these bonds.

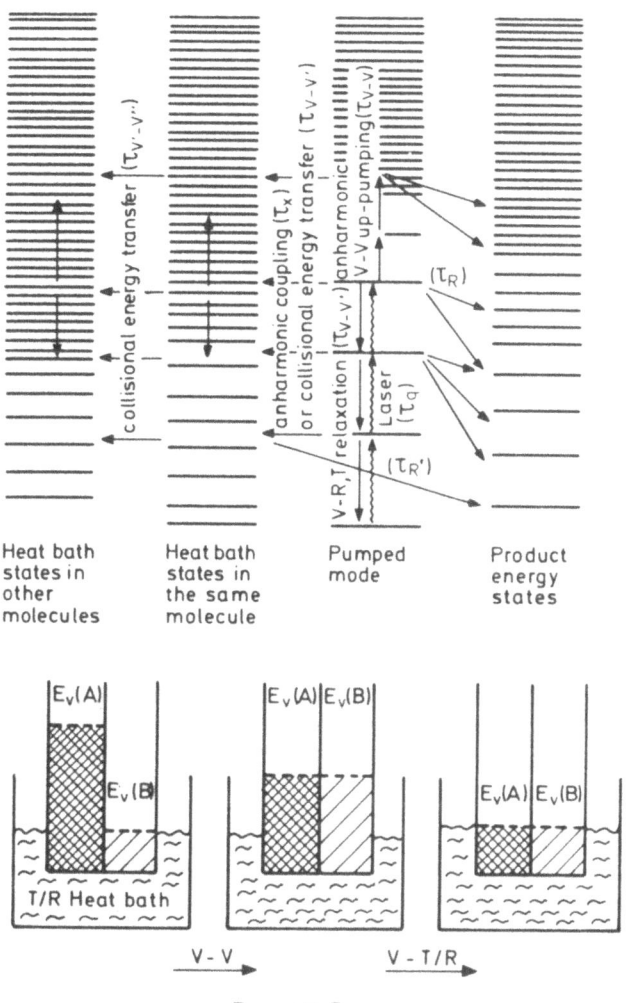

Fig. 4: Competing relaxation processes in a large molecule which
         may ruin the specificity of laser excitation. The ex-
         citation energy is leaking into the unimportant heat bath
         states in the excited molecule or in an admixture. The
         lower part of the figure shows that in this way the ini-
         tally high vibrational energy content in mode (A) is gra-
         dually shared with mode (B) by V⇋V transfer and finally
         with all the V/T/R states in the system. This implies a
         homogeneous temperature giving rise to thermal rather than
         laser-induced chemistry at the end.

there are striking examples of laser induced processes in which
the emission from reaction intermediates has been observed in the
process of falling apart.

In parallel to the analytical efforts and starting from a
different basis, practical chemists began using lasers in a rather
empirical way to induce chemical reactions. These groups utilized
mostly CO_2 lasers, pulsed and c.w., as this is the laser most
readily available to chemists. Also, because chemical reactions
generally involve rearrangements of nuclei in reactive systems,
vibrational photochemistry seems to be the most direct approach to
laser chemistry (Fig. 3). In this field many new reactions and
sometimes novel products were discovered without a complete understanding of the reaction mechanism. It is clear from the beginning
that the central problem in vibrational photochemistry is that of
energy localization. Likewise, the coupling strength between the
many vibrational degrees of freedom in a vibrationally excited molecule with more than just two atoms requires consideration (Fig.
4). This coupling - radiationless or collision-induced - will
eventually ruin the specificity of laser excitation and lead to a
homogeneous energy distribution in the molecules. It then becomes
a matter of timescales to intercept this relaxation trend by a
chemical reaction. The extent, to which this is possible is an
open question today.

A strong impact on laser chemistry has come through the dis-
covery of infrared laser-induced multiphoton excitation (dissociation, isomerization). This is not vibrational-mode selective, but
nevertheless a molecule selective process which can be employed to
separate trace constituents from a mixture. A very important example here concerns isotope separation of various elements throughout the periodic table. It has become customary to describe this
process of isotopically selective multiphoton dissociation (MPD)
by considering three regions of excitation (Fig. 5). The general
assumption then is that in the transition from region I to region
II all mode selectivity is irreversibly lost. This corresponds to
ergodic trajectories of the system under such conditions of excitation. Deviations from this behaviour are conceivable but may be
rarely occurring. For practical laser chemistry it is noteworthy
that with IR lasers large amounts of energy can be deposited in a
molecule, much more than would be expected from the resonant absorption provided by the molecular spectra. Thus very high vibrational temperatures can be obtained by this technique, perhaps
opening the way to new high temperature reactions. Studies of all
the various aspects of MPD have spread out enormously and cannot
be summarized here to any extent.

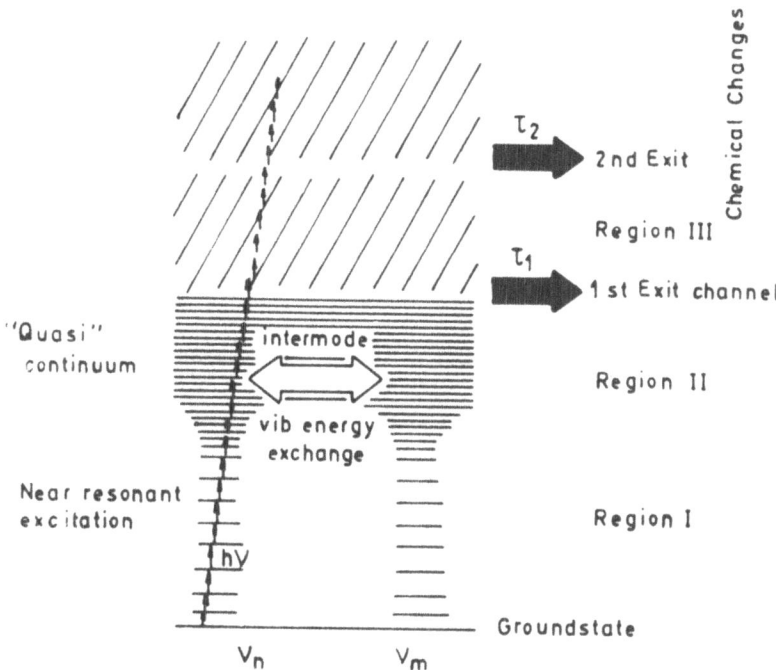

Fig. 5: Rough scheme of IR multiphoton excitation/dissociation.
In region I one deals with discrete energy levels; coherent excitation is possible, bottlenecking is expected
due to anharmonic shifting of the vibrational transitions
and isotopic (or species) selectivity can be achieved. In
region II intermode coupling becomes increasingly dominant and in region III (true continuum) chemical conversions are seen to occur.

Over the last years new high power lasers in the ultraviolet
spectral range have become available. Mainly the group of excimer/
exciplex lasers is to be mentioned here. These sources have consequently been applied to multiphoton excitation studies, too. For
polyatomic molecules UV laser-induced excitation very often results in ionization occurring after the absorption of two or three
UV photons (Fig. 6). Since the primary ions, however, may exhibit
UV (and visible) absorptions they tend again to undergo secondary
photochemical changes (Fig. 7). The global features of this type
of photoionization can be summarized as follows:

(I) Depending on laser parameters, fragmentation patterns can differ markedly from other ionization techniques (e.g. single photon photoionization, photoelectron spectroscopy, chemical- or charge transfer activation, electron impact). Fragmentations observed by MPI and other methods are linked; however, much higher energies can be deposited. Energy is the main controlling factor, no laser frequency dependence is expected as long as species absorb.

(II) Total ion yields can be very high (for resonant excitation approaching 100 %) allowing for very sensitive trace detection. Spectral dependence of ionization probability is allowing for selective detection of molecules ("Two-dimensional mass spectrometry"). Total ionization probability is dependent on intermediate state lifetimes.

(III) A model with multiple fragmentation steps is a good descriptions for many cases; support for statistical content of this model is provided by experimental data. Higher (but possible) excitation rates may lead to non-statistical behavior.

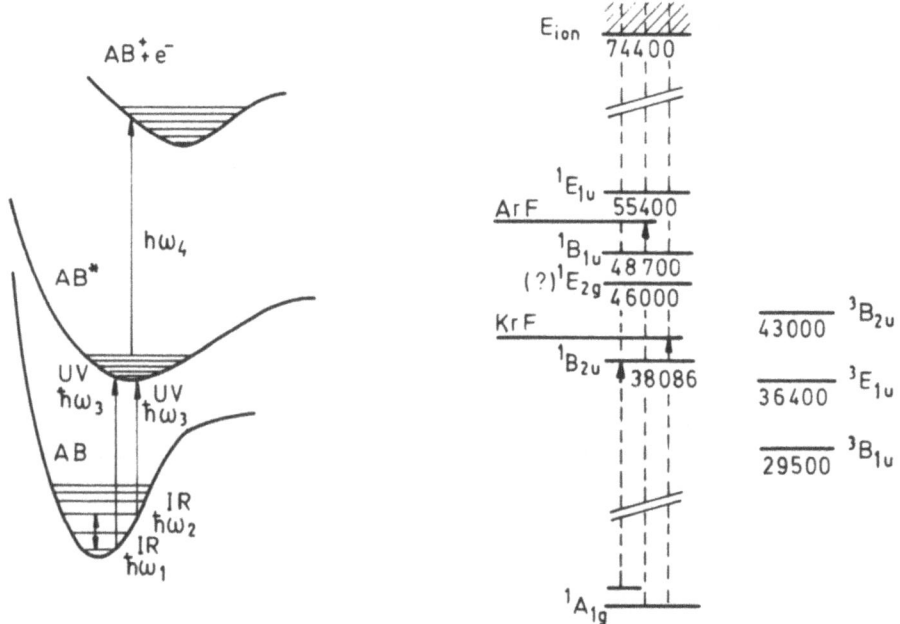

Fig. 6: Resonance enhanced 2-photon ionization shown for a general case (left) and for the specific case of benzene C_6H_6 (right, two different laser types).

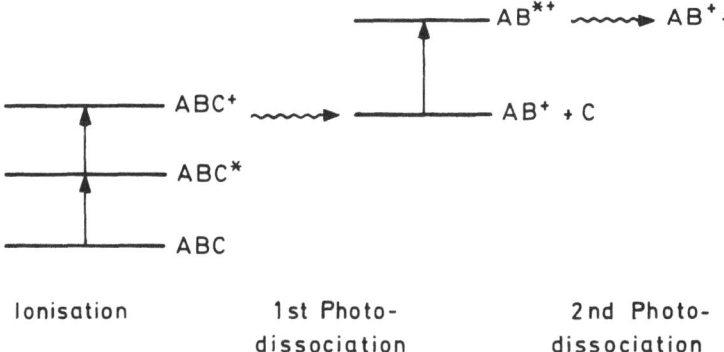

Fig. 7: "Ladder switching" mechanism for UV laser induced multiphoton ionization and fragmentation. Since the process
of photodissociation is not infinitely fast the sequence
of absorption/fragmentation steps can be interrupted
after the first step if a very short-pulse laser is used.

Table 1: Systematics of laser chemistry in relation to the most
important laser features.

Laser Source Property	Types of Applications	Typical Examples
Monochromaticity Tunability	State Selective Excitation / Probing	Basic Chemical Dynamics (Beams, Small Molecules) Laser Assisted Collisions
High Power / Energy	Multiphoton Excitation, High Excitation Rates	Separation, Purification (Isotopes) Controlled Radical For- mation, Photoionization
Short Pulses	High Temporal Resolution	psec Phenomena (Condensed Phases, Biophysics)
Collimation Focusability	High Spatial Resolution, Remote Heat Transfer	High Temperature Chemistry - Thermal Process Control, Microchemistry

Laser-induced photoionization is intimately related to molecular parameters like existence of resonant intermediate states, magnitude of ionization potential, occurrence of destructive processes affecting intermediate states and magnitude of transition probabilities.

It is thus possible to obtain information on these parameters through photoionization studies. The technique has found applications in mass spectrometry as a very efficient and (partly) selective ion source, and in ionization spectroscopy. As in the analogous case of IR multiphoton excitation, deviations from equilibrium energy distributions are conceivable or may even be more likely, but no systematic understanding is available as yet.

There are many diversified uses of lasers in chemistry and chemical physics. As returning again to the practice of laser chemistry, the following topics can be considered as receiving perhaps the most attention in the present scientific discussion.

In attempting a systematic view one may go either by the relevant laser properties (Table 1) or in a more empirical approach by the most discussed results as they appear in the literature (Table 2). Thus Table 1 and 2 may serve as an orientation and introduction to the chapters of this book.

Table 2: Empirical view of laser chemistry listing some of the most discussed topics.

Analytic	Synthetic	
(Static and Dynamic Systems)	(Thermal and Non-Thermal)	
	IR	UV
State to State Dynamics	Isomers	"Conventional"
Atmospheric Pollutants	Isotopes	Photochemistry
Combustion Monitoring	Purifi-	Radicals
Process Control	cation	
Surface Analysis	Radicals, Ions	
	Surface-	
	Treatment	

A final remark should be made with regard to the future development of the field. As it appears now there is no broad and general breakthrough in chemistry due to the use of lasers. There are however many scattered, yet often important applications in

chemistry. There is also a transgression from chemistry to materials research to be noted in the literature. This primarily concerns the study of surface processes and surface modifications
with lasers extending to topics like molecular beam scattering,
adsorption, desorption, catalytic reactions, oxidation, etching,
annealing, phase transitions, surface mobility, thin films, vapor
phase deposition and in general laser spectroscopy near an interface.

Considering these perspectives the future looks bright for
laser chemistry.

PART I

LASER AND RELATED LIGHT SOURCES

LASER SOURCES FOR CHEMICAL EXPERIMENTS

K.L. Kompa

Max-Planck-Institut für Quantenoptik
8046 Garching, Germany

The following brief review is intended to bridge the gap between laser physics and chemistry by summarizing the state of development of some laser relevant sources and also indicating some recent results with regard to energy output and frequency tunability.

For the chemical practitioner a laser is a delicate and expensive source of very high quality energy which allows him [1]
- to induce selective activation of molecules
- to engage in multiphoton and other novel photochemical concepts and
- to exert very detailed temporal and spatial control over reaction systems.

These objectives are linked to the field of laser spectroscopy. Indeed laser chemistry was born out of spectroscopy. On the other hand the more recent developments and trends in laser chemistry have characteristically diverged from this origin. This is particularly obvious for chemical experiments in condensed phases and on surfaces. Thus, as far as the laser requirements are concerned chemistry often requires less perfect spectral resolution than laser spectroscopy but more time resolution and generally higher laser powers and energies to effectively induce substantial molecular transformations. This will become apparent in the following.

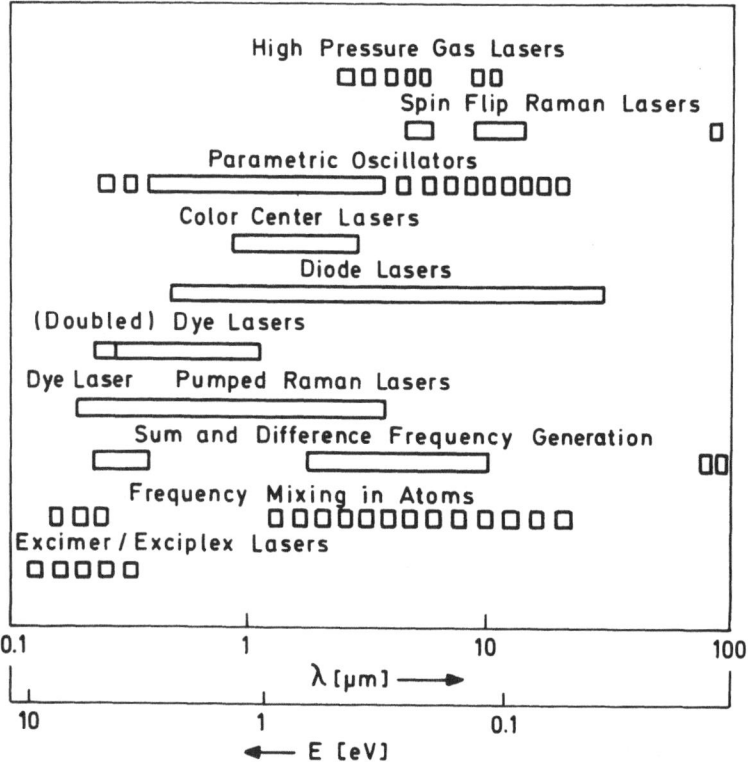

Fig. 1: Survey of the more important types of laser sources [2].
Out of this range mostly three groups - indicated in
the figure - have been used in practice for chemical
experiments.

The following discussion is intended to contribute to synthetic uses of lasers. The equally important range of diagnostic
and analytic laser applications is therefore not specifically
covered [3]. The general understanding here is that for analytical
chemistry most of the laser sources needed are available while it
is synthetic chemistry where there is still a lack of suitable
laser technology.

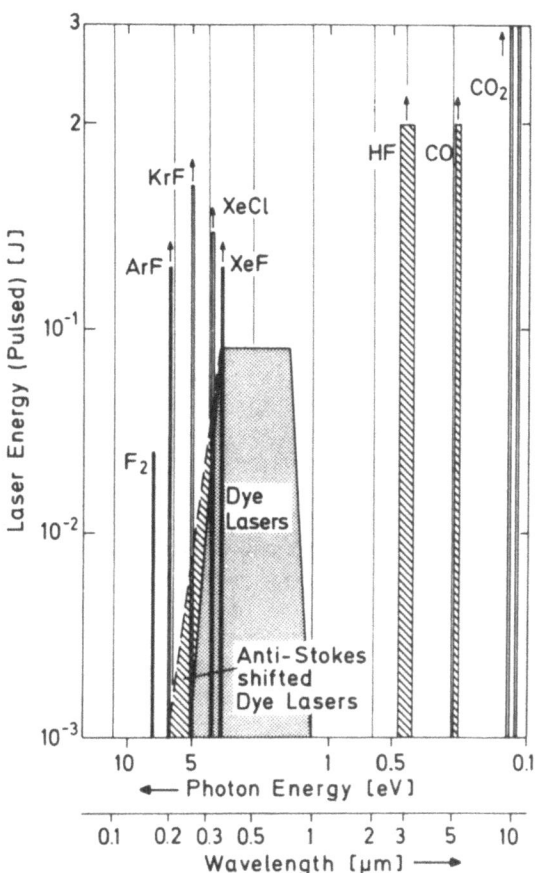

Fig. 2: Wavelengths and typical pulse energies of some lasers of relevance to chemistry.

There is a tremendous number and variety of laser sources as Fig. 1 shows. Among them mostly three groups have gained importance for chemical applications, namely the molecular gas lasers, notably CO_2, in the infrared, the wellknown dye lasers in the visible and the relatively new group of UV excimer lasers. Thus Fig. 2 summarizes wavelengths and typical output energies of these three groups of lasers.

A systematic discussion of the advantages of lasers show that these light sources are superior to conventional lamps by their higher intensity, better monochromaticity, higher quantum fluence per frequency interval, better divergence or beam collimation, their polarisation characteristcs and by their short and ultrashort pulses. Especially the combination of several of these features in a single laser source appears noteworthy.

A point of special relevance to practical chemistry is the energy output of the laser because this is the parameter which determines the absolute amount of photochemical product. This energy question shall first be considered for the case of a UV laser with 5 eV/Photon (250 nm, e.g. KrF laser). To obtain an analyzable amount of photochemical reaction product (~2 mg pure substance for e.g. NMR or chromatographic analysis) a total input of laser energy of ~ 10^5 J (if the quantum yield is as low as $\phi = 10^{-3}$, molecular weight of product assumed to be between 100 and 200), ~ 20 J (if $\phi=1$) or ~ 0.1 J (if $\phi=10^3$) is required.

This implies for the least favorable case mentioned ($\phi ~ 10^{-3}$) that one such experiment would indeed exhaust the usable lifespan of a present day excimer laser (see below) and would therefore be considered impractical. For a high yield product, on the other hand, the situation looks quite favorable from this point of view. No such problems exist for the corresponding case of infrared laser photochemical experiments at least as long as the most developed CO_2 or CO lasers are employed. Other infrared and visible laser sources need special consideration.

Another most important laser feature in the present context is general frequency tunability. This is a less stringent requirement than in spectroscopic laser applications; nevertheless most reagents absorption features are discrete (bandwidth ~ 0.1 - 10 cm^{-1}) and hence different wavelengths can obviously have different effects on reaction systems. As Fig. 2 shows only dye lasers in the visible spectral range exhibit perfect tunability [3]. For the IR as well as the UV range this question requires some more detailed discussion.

The available molecular gas lasers in the infrared all of

which can be operated both pulsed and cw are shown again in Fig. 3.
(comp. also Fig. 1) The frequency ranges indicated are covered
intermittently by vibrational-rotational emission lines of the
lasing molecules. Continuous tunability can in principle be achieved by pressure broadening. For a CO_2 laser this requires 20 bar
pressure and the corresponding measures to generate a stable
electrical discharge under these conditions (e.g. e-beam pre-
ionization). Fig. 4 shows the performance of such a system [4].
It appears particularly noteworthy that wideband emission
($\Delta\upsilon \sim 3$ cm^{-1}) or likewise simultaneous emssion at different

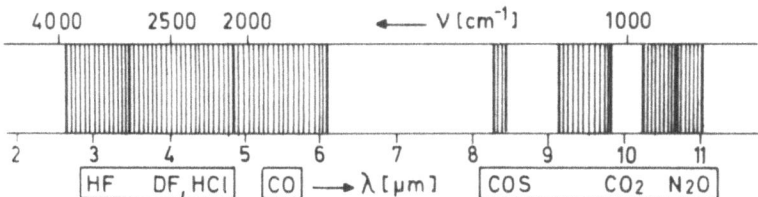

Fig. 3: Some prominent infrared molecular gas lasers capable of
        high power operations.

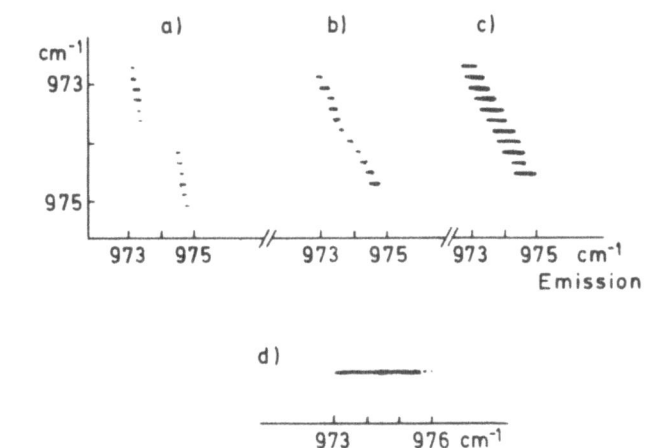

Fig. 4: Emission of high pressure CO_2 laser as function of grating
        position at a) 10, b) 15, c) 20 bar. d) broadband emission
        without grating [4].

wavelengths is possible. This may serve as an example to indicate only roughly some of the design parameters and development goals with regard to IR lasers suited for chemical applications.

In the UV range the situation is different. Again we begin our discussion with a survey of the most popular excimer lasers, Fig. 5 [5]. The spectroscopic level scheme of these lasers is such that a group of bound upper states is connected to a weakly bound or repulsive ground state. This is shown for the general type of excimer lasers and also for the somewhat special case of the halogen lasers [6] in Fig. 6 but applies with modifications to all these lasers. Correspondingly the fluorescence on this transition is more or less broad and unstructerred and gives hope for some tunability similar to the case of dye lasers. Depending on the special excimer laser system in question this may in principle be utilized. However for practical reasons it may be more convenient to use these excimer systems only as amplifiers rather than oscillators. The tunable radiation to control such an amplifier may then be obtained by high-order Anti-Stokes shifting of an - again excimer laser pumped - tunable dye laser into the short wavelength range (Fig. 7)[7]. Without any detailed discussions and only referring to users interests Fig. 8 exemplifies the tuning range for AS3 and AS7. The latter lies around the wavelength of the ArF excimer laser. The problem then remains now to couple such

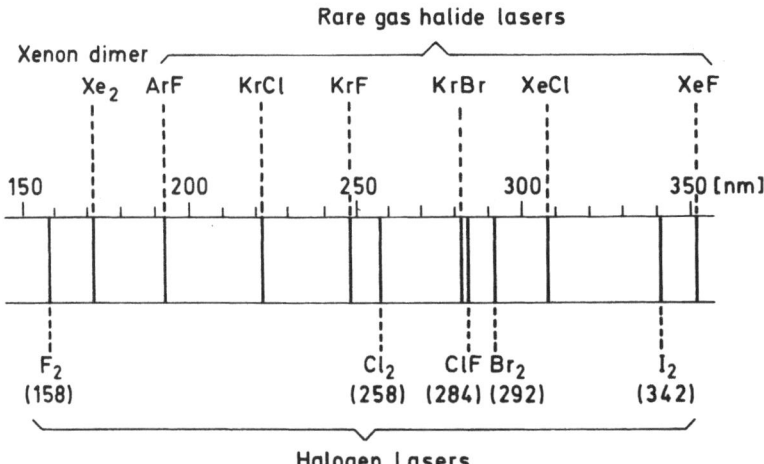

Fig. 5: Survey of the most developed rare gas excimer, rare gas halide and diatomic halogen lasers (potential tuning ranges not shown only centre wavelengths of emission given).

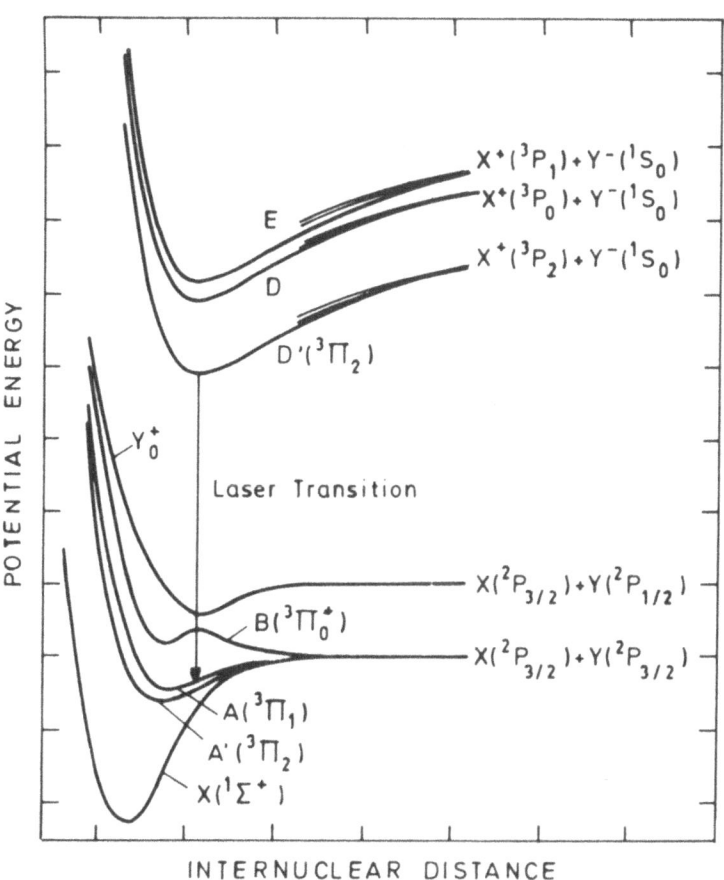

Fig. 6: a) Assumed spectroscopic level scheme of halogen lasers[6].

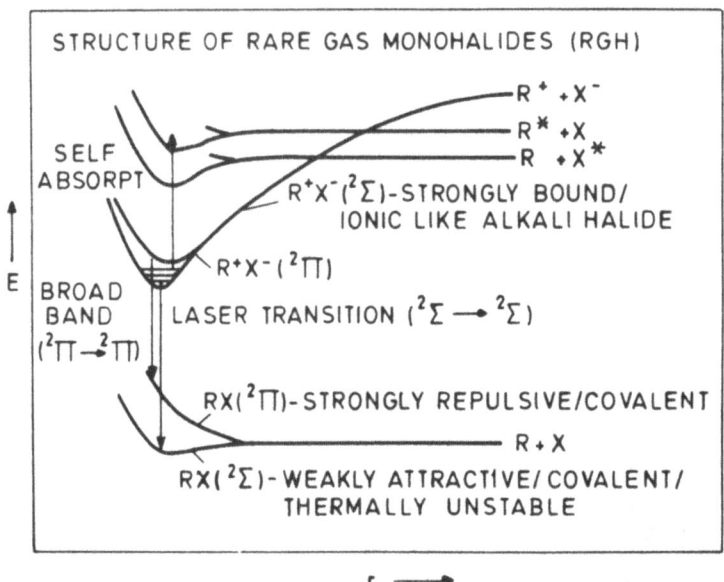

Fig. 6: b) Level structure of rare gas halide exciplex lasers[5].

a source to the amplifier. The success of this scheme will depend
on a judicious balance of timing and synchronization (often multiple passes through the amplifer will be advantageous) as well
as of adjusting amplifier gain to the appropriate level. It
should be mentioned that this type of Raman-shifted oscillator
plus amplifier arrangement is obviously not the only way to tunable high power UV lasers but can be considered as a comparatively simple scheme. The present state of development of excimer
lasers still leaves much to be desired but the principal potential for new photochemical sources should be obvious from Figs. 2
and 6-8.

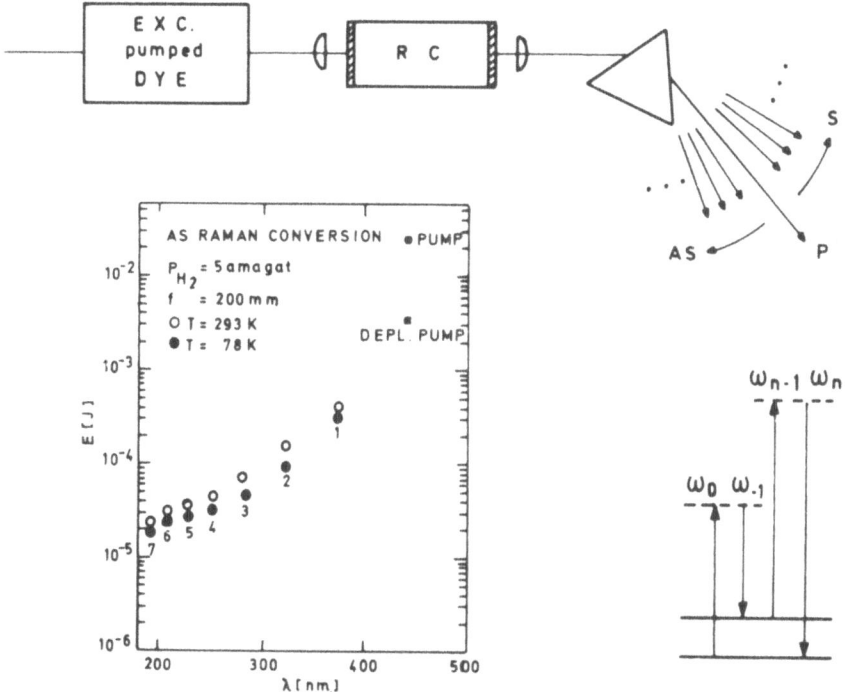

Fig.7: Physical principle experimental set-up and results for AS raman conversion of coumarin 2 dye laser [7].

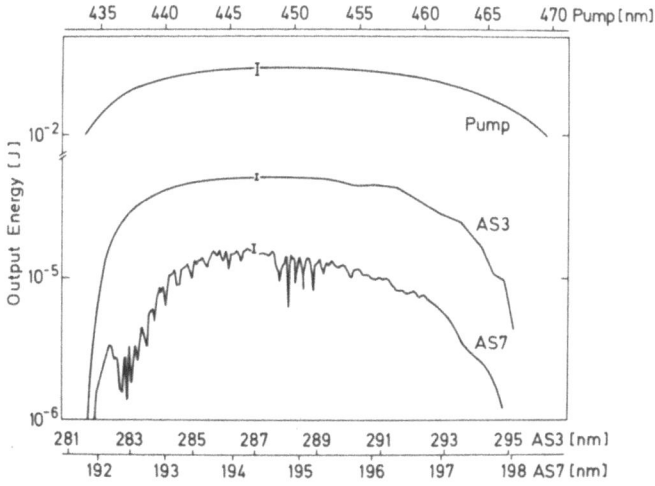

Fig. 8: Tuning ranges of AS_3 and AS_7 with coumarin 2 at room temperature [7].

REFERENCES

1   Springer Series in Chemical Physics, Springer Verlag
    Berlin-Heidelberg-New York
    Vol. 3 Advances in Laser Chemistry, Ed. A.H. Zewail
    Vol. 10 Lasers and Chemical Change,
        A. Ben-Shaul, Y. Haas, K.L. Kompa, R.D. Levine
    Vol. 22 Nonlinear Laser Chemistry Multiple-Photon Excitation,
        V.S. Letokhov
    Chemical and Biochemical Applications of Lasers,
    Ed.C. B. Moore, Vol. I-V, Academic Press New York
    Megawatt Infrared Laser Chemistry, E. Grunwald, D.F. Dever,
    P.M. Keehn, John Wiley, New York
    Laser-Induced Chemical Processes, Ed. J.I. Steinfeld
    Plenum Press, New York
2   Laser Handbook, North Holland, Amsterdam, Vol. 1-2, Ed.
    F.T. Arecchi, E.O. Schulz-Dubois, Vol. 3, Ed. M.L. Stitch
    Handbook of Chemical Lasers, Ed. R.W.F. Gross, J.f. Bott
    John Wiley, New York
    High-Power Lasers and Applications, Ed. K.L. Kompa and
    H. Walther, Springer-Verlag, Berlin-Heidelberg-New York
    Electronic Transition Lasers, Ed. J.I. Steinfeld, The MIT
    Press, Cambridge, Mass.
    Electronic Transition Lasers II, Ed. L.E. Wilson, S.N. Suchard,
    and J.I. Steinfeld. MIT Press, Cambridge, Mass.
3   Dye Lasers, Ed. F.P. Schäfer, Springer-Verlag,
    Berlin-Heidelberg-New York
4   Wan Chong-Yi, C. Schwab, W. Fuß and K.L. Kompa
    Opt. Comm. in print 1983
5   Excimer Lasers, Ed. C.K. Rhodes, Springer-Verlag
    Berlin-Heidelberg-New York
6   M. Diegelmann, F. Rebentrost, K.L. Kompa,
    J. Chem. Phys. (1982) $\underline{76}$, 1233
7   D.J. Brink and D. Proch, Opt. Lett. $\underline{7}$, 494 (1982)
    D.J. Brink, D. Proch, D. Basting, K. Hohla, P. Lokai,
    Laser und Optoelektronik, Nr. 3/1982

HIGH POWER OPTICALLY PUMPED MID-INFRARED

MOLECULAR GAS LASERS

R.G. Harrison

Physics Department
Heriot-Watt University
Riccarton, Currie, Edinburgh, EH14 4AS

INTRODUCTION

The search for efficient and powerful sources of coherent radiation in the mid-infrared (MIR) spectral region of 5-20 μm has in the last few years been particularly motivated by the need of these sources for laser photo-chemical reactions, an important one of which is selective multiple photon dissociation of uranium hexaflouride. Of the various schemes optical pumping of molecular gases[1-4] has proven the most successful approach providing an efficient, non-destructable and simple method for infra-red generation adaptable to a large variety of molecules and therefore particularly useful in the search for new MIR laser emissions. Based on the resonant excitation of vibrational rotational transitions, for which moderately high gains of 10^{-2} to 10^{-1} cm^{-1} may be readily obtained, many emissions in the spectral region have already been obtained (see fig. 1) with powers in favourable cases being in excess of 1 MW and photon conversion efficiencies approaching 100%.

The striking success of this approach owes much to the development of several powerful pulsed discharge excited molecular lasers. Of these the TEA CO_2 system has emerged as the dominant pumping source providing an efficient multi megawatt single line emission step tunable (~ 2 cm^{-1} step interval) over a fairly wide range (9-11 μm) for which numerous molecules have vibrational absorption bands. The emissions show in fig. (1), obtained using this pump source, contribute most of those reported over the last six years; the predominance of emissions around 16 μm being in response to the demand for a source at this wavelength for the uranium enrichment programme.

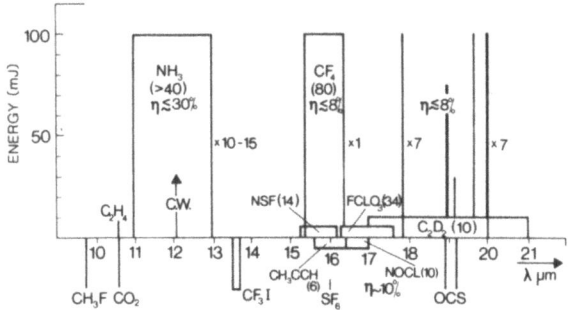

Fig. 1 OPML emissions
obtained with TEA CO_2
laser. Those below axis
have unspecified output
energies. Number of
emissions (in brackets)
together with efficiency
η are given for each
gas.

SPECIFIC OPTICALLY PUMPED MOLECULAR SYSTEMS

Depending on the nature of pumping and lasing transitions
OPMLs can be categorised in 3 groups as shown in fig. 2, each of
which has its own characteristic features.

a) Lasers based on inversion in a hot band

For the first subgroup, type A_1, a combination level is
excited and lasing occurs on a hot band. Type A_2, where an overtone level is excited followed by lasing on a single vibrational
quantum can be considered to be a special case of type A_1. In
both of these, because the pump transition involves two or more
vibrational quanta absorption of the pump beam is often small, but
gain on laser transitions can be large since a single quantum
transition occurs.

These lasers are far the most promising as regards a search
for new MIR lasers, most of the emissions shown in fig. (1) being
of this type. However pressure scaling of these systems,
desirable for continuously tunable emission is limited since
collisional depopulation of the excited levels can be fast and
also because of the presence of ground state absorption close to
the frequency of the lasing band.

Of the large number of molecules that have been made to lase
in the 16 μm region, namely CF_4[5], $NOC\ell$[5], $FC\ell O_3$[6], NSF[7], CH_3CCH[8],
NH_3[9-11] and CO_2[12] the most extensively developed is CF_4 for which
outputs as high as 100 mJ with photon conversion efficiencies of
$\sim$ 5% have been reported by several authors[13,14]. More than 80
emissions around 615 cm^{-1} have been obtained using different
isotope forms of the pump and lasing molecule[15] and the development
of high pulse repetition rate systems (80 Hz) have yielded average

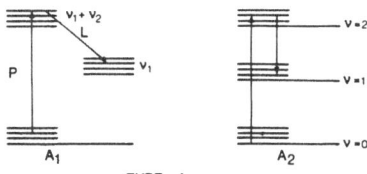

TYPE A

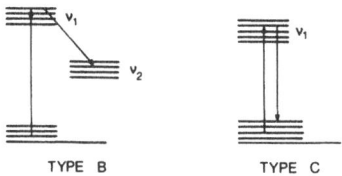

TYPE B   TYPE C

Fig. 2. Energy level schemes for mid-infrared OPML's.

powers of 2.5W [16]. A common feature of lasers of this type is the significant improvement in the maximum operating pressure (for CF_4 from 5 to 23 torr) and energy output achieved on cooling the gas[17] thereby reducing the population of the terminal level of the lasing transition and also enabling optimization of the population distribution of the lower pump level. Multiphoton dissociation of UF_6 has been studied using the CF_4 system[18,19]. Emissions from $FC\ell O_3$ [6] in 565-613 cm^{-1} region, from NSF[7] in 618-650 cm^{-1}, and from propyne[8] in 609-637 cm^{-1} range compare favourably in performance to that of CF_4 and although as yet not fully developed have the added advantage of spectral features in better match with those desirable for laser isotope seperation of Uranium.

Notable at longer wavelengths is the impressive energy extractions (as high as 750 mJ) from some of the fifteen emission lines in the 17.4 - 20.5 µm region from C_2D_2[20,21]. With an HF laser pump hot band lasing has been obtained over a number of lines in 4.4 to 18 µ region using $^{15}N_2O$, $^{14}N^{15}NO$, $^{15}N^{14}NO$, HCOOH,[13] CS_2 (22), $^{15}NH_3$ (23) and $^{12}C^{18}O$, $^{12}C^{16}O^{18}O$, $^{13}C^{16}O^{18}O$ (24) molecules. Hot band lasing at 16 µm and difference band at 14 µm was also obtained[12] using an HBr laser to pump 1:1 mixture of HBr and CO_2.

Laser emissions following two photon excitation have been reported from SF_6[25], NH_3[9-11], and CH_3F[26], most notable of which were the 10 emissions from NH_3 in the 6-35 µm region with energy extraction of a few milli joules following excitation of the 2a ν_2(5,4) level of the molecule. Other hot band lasers include OCS[27,28] (type A_2), CF_3I[29] and COF_2 [2].

b) <u>Lasers based on inversion in a difference band</u>

        In lasers belonging to group B a fundamental level is pumped
and laser action occurs within a difference band to a lower level.
Here the pump absorption cross section can be large, although,
since difference band transitions are forbidden in the harmonic
approximation, the gains of these systems can be small.   Very
high pressure operation is however possible for this type of laser
due to the absence of significant self absorption and the long
collisional relaxation times associated with fundamental molecular
modes.   Two molecules that have been made to lase efficiently
on this scheme are CO_2 and N_2O.   In both these cases very high
pressure operation (30-40 atm) was achieved permitting generation
of continuous tunable MIR output.   For CO_2[30,31] the 4.3 μm line
of HBr was used to pump the ν_3 band of CO_2 resulting in complete
conversion of 4.23 μm photons to 10.6 μm photons.   Due to the
small penetration depth of the pump beam resonator lengths as
short as 1 mm were used similar to that for the N_2O system[31,32]
for which lasing occured via resonant collisional energy transfer
to N_2O from CO_2 excited by an HBr laser.   With this system continuous tuning of over 5 cm^{-1} near 10.5 μm was obtained with a
resolution of 0.014 cm^{-1}.

        The versatility of collisional energy transfer has also been
demonstrated using v-v collisional transfer from CO, excited by
the second harmonic of a CO_2 laser line, to obtain MIR emission
in the difference band of 6 molecules[33].   The CO molecule is
ideal for storage of vibrational energy because of its exceedingly
slow vibration to translation transfer rate of 1.9 x 10^{-1} sec^{-1}
$torr^{-1}$.   It is also efficiently excited by the second harmonic of
the CO_2 9P(24) line which falls within 0.003 cm^{-1} of the CO 0→1
P(14) transition.   Of the lasers, OCS-CO was notable for grating
tuned emission over 80 lines between 8.19 and 8.46 μm with maximum
outputs of 1.3 mJ and efficiencies of ∿ 7% to be compared to 19%
for direct pumping (2ω CO_2) of OCS.

c) <u>Lasers based on fundamental bands</u>

        For this laser (Fig. 2 type C), the pump and lasing occur on
different lines within the same fundamental band.   For example
resonant excitation of the R-branch line, can be followed by
lasing on a P branch line.   The cross section for both the pump
and lasing transitions can be large in this case resulting in
very efficient laser performance.   Simple spectroscopic considerations show that this scheme is limited in application to light
molecules for which the rotational level spacing is sufficiently
large to permit gain in the presence of the thermal population of
the terminal level, although even here the available transitions
are found to be limited.   The most important of the molecules

suitable to this scheme is NH_3[34] emitting in the 10.7 to 13.9 μm region[35,15].   Power conversion efficiencies of 40-80%[36] and energy extractions in excess of 1 J with efficiencies of > 20% have been obtained on some of the 50 emission lines[37], making this system the best of the OPML's to date.   Many of the emissions have been obtained with the NH_3 buffered by large amounts of N_2, so ensuring rapid rotational thermalization in all the vibrational states and also depletion of the population of the terminal lasing states by resonant energy transfer to levels of N_2, resulting in population inversion for most of the vibrational-rotational trans-itions whose frequencies are lower than that of the pump[37,38]. Other emissions are attributed to lasing among the rotational levels of the ν_2 vibrationally excited mode involving far infrared emission and subsequent MIR generation[39].   Several emissions obtained in which both the pump and lasing frequencies are off-resonant, are attributed to Raman like processes[40]. - To date the NH_3 OPML is the only MIR system for which c.w. action has been obtained for which an output of 180 mW was generated at 12.08 μm[41].

Since the first report of the NH_3 laser[34] the system has received considerable attention due to its applications in tunable mid-infrared generation by nonlinear mixing techniques[42-44] and as pump for InSb Spin Flip Raman (SFR) laser[45] and in laser photo-chemistry[46,47].

EXPERIMENTAL CONSIDERATIONS

The optimum performance of an OPML poses the simultaneous requirement of injecting a high power pump laser beam into an absorbing gas while maintaining a high Q cavity at the pumped laser wavelength.   Early systems[34] utilizing off axis injection of the pump beam in the OPML cavity resulted in critical and in-efficient operation, chiefly due to the nonuniformity of pumping which can result in self absorption of stimulated emission by the unexcited gas.   This problem is minimized by having the CO_2 pump propogate collinear to the axis of the optically pumped cavity, using for example specifically coated dielectric mirrors and filters to transmit the CO_2 pump and form the OPML cavity[48], although such schemes are prone to optical damage.   Subsequent schemes have used cavity configurations with damage resistant original gratings, a particularly versatile and simple one[36]   being shown in Fig. 3.   Based on the use of a common output coupler for both the CO_2 laser and the OPML (the gains of which are similar) the beams are maintained collinear and the output is efficiently coupled directly off the TEA CO_2 intra-cavity grating.

In overcoming the limitations to pump energy inherent to the conventional stable resonator configurations commonly used for TEA CO_2 lasers alternative systems have been used and included oscillator-amplifiers[13] and unstable resonator cavities[49].

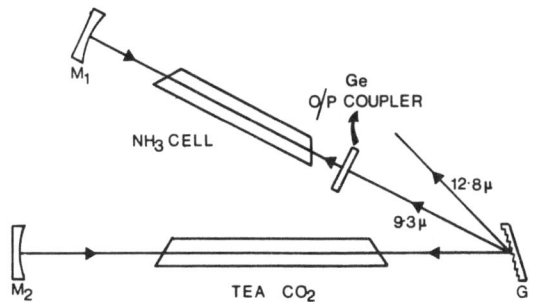

Fig. 3. Pumping and cavity
arrangement using
a common output
coupler.

Attention has also been given to operation of the pump laser
on a single mode thereby ensuring greater amplitude stability and
improved coupling of the pump radiation to the resonant transition
of the pumped gas[50,51]. The conventional multi mode emission
from TEA CO_2 lasers spans a bandwidth of 1-2 GHz often significantly
greater than the OPML transition to be pumped.

For OPML's that operate at reduced temperature (in particular)
Type A systems), gaseous boil off from liquid nitrogen is commonly
used to cool the OPML cell. Improved design of cryogenic systems
now ensures temperature stability of 1.5°K and uniformity over
many hours[51].

Optoacoustic technique, commonly used in the past for FIR
lasers are now also routinely used for the MIR for finding spectral
coincidences between pump lasers and candidate gases.

THEORETICAL CONSIDERATIONS

The gain for OPML may be due to either coherent Raman
type processes or to pure population inversion effects or to both.
For a complete description of the situation a quantum mechanical
approach is required as has been used for FIR lasers[52,53].
However, in situations where the coherent effects are not important
i.e. gain is purely due to population inversion, a rate equation
approach is adequate to describe the basic gain features of these
lasers[4,54,55]. This approach is justified to the extent that the
pump sources are generally multimode lasers and the excited molecules are at pressures for which collisional energy transfer in-
variably plays a significant role.

The development of a theoretical basis for the operating
characteristics of optically pumped mid-infrared gas lasers is
still in an embroyonic stage. Thus so far only two attempts to
model a mid-infrared OPML have been reported, the molecules being

OCS^{28} and $NH_3{}^{56}$ for both of which adequate information about their spectroscopic and energy transfer aspects were available. Both the attempts have used the rate equation approach and result in good agreement between experimental and theoretical data.

CONCLUSIONS

The recent progress in optically pumped molecular lasers (in particular those using CO_2 laser pump) has proved their potential as an efficient source of high power MIR radiation. The development of OPMLs has so far been dominated by the demand for a coherent source suitable for uranium enrichment programme as is evidenced from Fig. 1. Recognising the urgent need for extending the spectral coverage of such sources for other applications, in particular their more general use in infrared laser photochemistry, optical pumping of molecular gases by TEA CO_2 laser radiation will with similar effort undoubtedly become an important general source of powerful radiation over a large number of lines for a variety of different molecules in the MIR spectral range of 11-25 µm.

REFERENCES

1. T.Y. Chang, in Nonlinear Infrared Generation, Y.-R. Shen, ed. Springer-Verlag, Berlin (1977), pp215-272.
2. C.R. Jones, Laser Focus, 14, No. 8, 68 (1978).
3. A.Z. Grasiuk, V.S. Letokhov, and V.V. Lobko, Prog. Quant. Electr. 6, 245, (1980).
4. R.G. Harrison, P.K. Gupta, in Infrared and Millimeter Waves, vol. 7, K.J. Button, ed. to be published by Academic Press.
5. J.J. Tiee and C. Wittig, Appl. Phys. Lett. 30, 420 (1977).
6. H.N. Rutt, Opt. Commun. 34, 434 (1980).
7. T.A. Fischer, J.J. Tiee and C. Wittig, Appl. Phys. Lett. 37, 592 (1980).
8. T.A. Fischer and C. Wittig, Appl. Phys. Lett. 39, 6 (1981).
9. R.R. Jacobs, D. Prosnitz, W.K. Bischel and C.K. Rhodes, Appl. Phys. Lett. 29, 710 (1976).
10. J. Eggleston, J. Dallarosa, W.K. Bischel, J. Bokor and C.K. Rhodes, J. Appl. Phys. 50, 3867 (1979).
11. A.N. Bobrovskii, A.A. Vedenov, A.V. Kozhevnikov and D.N. Sobolenko, JETP Lett. 29, 537 (1979).
12. R.M. Osgood Jr., Appl. Phys. Lett. 32 564 (1978).
13. J.J. Tiee, T.A. Fischer and C. Wittig, Rev. Sci. Instrum. 50, 958 (1979).
14. V. Yu Baranov, S.A. Kazakov, V.S. Mezhevov, A.N. Napartovich M. Yu Orlov, V.D. Pismennyi, A.I. Starodubtsev and A.N. Startostin, (1980), Sov, J. Quantum Electron 10, 47.
15. Handbook of Laser Science and Technology, vol. II, M.J. Weber, ed., to be published by CRC Press, Boca Raton, EL.

16. V. Baranov, B.I. Vasilev, E.P. Velikhov, Yu. A. Gorokhov, A.Z. Grasyuk, A.P. Dyad'kin, S.A. Kazakov, V.S. Lotokhov, V.D. Pismenny and A.I. Starodubtsev, Sov. J. Quantum Electron 8, 544 (1978).
17. J.M. Green, J. Phys. D: Applied Physics 12, 489 (1979).
18. J.J. Tiee and C. Wittig, Opt. Commun. 27, 377 (1978).
19. J.A. Horsley, P. Rabinowitz, A. Stein, D.M. Cos, R.O. Brickman and A. Kaldor, IEEE J. Quantum Electron. QE-16, 412 (1980).
20. H.N. Rutt and J.M. Green, Opt. Commun. 26, 422 (1978).
21. B.K. Deka, P.E.Dyer and R.J. Winfield, Opt. Lett. 5, 194,(1980).
22. A.H. Bushnell, C.R. Jones, M.I. Buchwald and M. Gundersen, IEEE J. Quantum Electron. QE-15, 208 (1979).
23. C.R. Jones, M.I. Buchwald, M. Gundersen and A.H. Bushnell, Opt. Commun. 24, 27 (1978).
24. M.I. Buchwald, C.R. Jones, H.R. Fetterman and H.R. Schlossberg, Appl. Phys. Lett. 29, 300 (1976).
25. W.E. Barch, H.R. Fetterman and H.R. Schlossberg, Opt. Commun. 15, 358 (1975).
26. D. Prosnitz, R.R. Jacobs, W.K. Bischel and C.K. Rhodes, Appl. Phys. Lett. 32, 221 (1978).
27. H.R. Schlossberg and H.R. Fetterman, Appl. Phys. Lett. 26, 316 (1975).
28. E. Armandillo and J.M. Green, J. Phys. D: Appl. Phys. 11, 421 (1978).
29. J.J. Tiee and C. Wittig, J. Appl. Phys. 49, 61 (1978).
30. T.Y. Chang and O.R. Wood II, Appl. Phys. Lett. 23, 370, (1973).
31. T.Y. Chang and O.R. Wood II, IEEE J. Quantum Electron. QE-13, 907 (1977).
32. T.Y. Chang and O.R. Wood II, Appl. Phys. Lett. 24, 182 (1974).
33. H. Kildal and T.F. Deutsch, Appl. Phys. Lett. 27, 500 (1975).
34. T.Y. Chang and J.D. McGee, Appl. Phys. Lett. 28, 526 (1976).
35. S.M. Fry, Opt. Commun. 19, 320 (1976).
36. P.K. Gupta, A.K. Kar, M.R.Taghizadeh and R.G. Harrison, Appl. Phys. Lett. 39, 32 (1981).
37. B.I. Vasilev, A.Z. Grasyuk, A.P. Dyadkin, A.N. Sukhnov and A.B. Yastrebkov, Sov. J. Quantum Electron 10, 64 (1980).
38. H. Tashiro, K. Suzuki, K. Toyoda and S. Namba, Appl. Phys. 21, 237 (1980).
39. T. Yoshida, N. Yamabayashi, K. Miyazaki and K. Fujisawa, Opt. Commun. 26, 410 (1978).
40. T.Y. Chang and J.D. McGee, Appl. Phys. Lett. 29, 725 (1976).
41. C. Rolland, B.K. Garside and J. Reid, Appl. Phys. Lett. 40, 655 (1982).
42. R.G. Harrison and F.A. Al-Watban, Opt. Commun. 20, 225 (1977).
43. R.G. Harrison, R.A. Wood and S.R. Butcher, Opt. Commun. 27, 157 (1978).
44. R.G. Harrison, P.K. Gupta, M.R. Taghizadeh and A.K. Kar, IEEE J. Quantum Electron QE-18, 1239, (1982).
45. C.K.N. Patel, T.Y. Chang and V.T. Nguyen, Appl. Phys. Lett. 28, 603 (1976).

46. R.V. Ambartsumian, Z.A. Grasiuk, A.P. Dyadkin, N.P. Furzikov, V.S. Letokhov and B.O. Vasil'ev, Appl. Phys. 15, 27 (1978).

47. J.J. Tiee and C. Wittig, Appl. Phys. Lett 32, 236 (1978).

48. B. Walker, G.W. Chantry and D.G. Moss, Opt. Commun. 23, 8, (1977).

49. B.K. Deka, P.E. Dyer and I.K. Perera, Opt. Commun. 32, 295 (1980a).

50. T. Stamatakis and J.M. Green Opt. Comm. 30, p413 (1979).

51. H.N. Rutt, (1981), Proceedings of Vth National Quantum Electronics Conference, Hull, Sept. 1981 (to be published).

52. T.A. DeTemple (1979) in Infrared and Millimeter Waves Vol. 1 ed. Button K, (Academic Press, N. York, San Francisco, London) p129.

53. T.Y. Chang, IEEE J. Quantum Electron QE-13, 937 (1977).

54. A.L. Golger and V.S. Letokhov Sov. J. Quantum Electron 3, 15, (1973).

55. A.L. Golger and V.S. Letokhov, Sov. J. Quantum Electron 3, 428, (1974).

56. P.K. Gupta and R.G. Harrison, IEEE J. Quantum Electron QE-17, 2238 (1981).

SYNCHROTRON RADIATION AND LASER EXCITATION SOURCES FOR STUDIES OF

INTRAMOLECULAR DYNAMICS

Sydney Leach

Laboratoire de Photophysique Moléculaire du C.N.R.S.
Bâtiment 213
Université Paris-Sud - Orsay Cedex, France

ABSTRACT

Ideal source requirements for selective excitation studies of intramolecular dynamics are discussed. The properties of synchrotron radiation are presented and are shown to be useful for such studies. A comparison is made between VUV laser and SR sources and their complementary nature for studies in the high energy region is shown.

I. INTRODUCTION

This lecture concerns selective excitation sources for studies of intramolecular (and intermolecular) dynamics. Selective excitation of a molecule requires optical excitation rather than (less selective) methods such as charged or neutral particle impact. The ideal source requirements are as follows[1] :

1) Tunability over a broad energy range (0.1-1000 eV). Figure 1 indicates the energy regions for excited state photophysical and photochemical processes in molecules as well as the present range of synchrotron and laser sources.

2) Time resolution. Time scales for intramolecular dynamics vary from 10^{-16} s (electronic autoionization), through $10^{-11} - 10^{-14}$ s (vibrational relaxation and vibrational predissociation) $10^{-4} - 10^{-13}$ s (electronic predissociation), to seconds (long-lived luminescence ; very slow electronic relaxation). Recent developments in laser technology, synchrotron radiation technology and modern

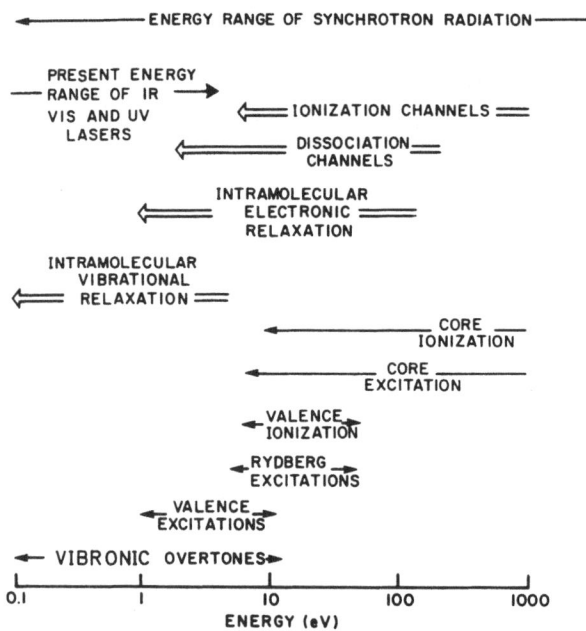

Fig. 1. Energy regions for excited state photophysical and chemical
        processes in molecules.

detection methods make it possible to conduct routine time-resolved
measurements on the nanosecond time scale. The lower limit for the
current "state of the art" is in the picosecond range for synchrotron radiation[2] and the femtosecond range for laser radiation[3].

3) Energy resolution. The spectral resolution for energy-resolved
observables depends on the type of physical information required.
Very high resolution (< 10^{-3} cm^{-1}) is required for the study of
slow relaxation of a single rotational level. For many molecular
relaxation processes, energy resolution in the 1 cm^{-1} range is
often sufficient, while for ultrafast processes 10-100 cm^{-1} is
usually adequate.

4) Intensity requirements. The intensity of the excitation source
should be sufficient for time-resolved and/or energy-resolved studies over the appropriate energy range. This in turn depends on
the detection methods and on whether single or multiphoton processes are examined.

5) High degree of collimation. Useful for increasing effective intensity (brightness) of excitation sources.

6) <u>Polarization</u> : linear or circular polarization is very useful for studying angular distribution in ionization and fragmentation processes as well as for studies of energy transfer in condensed phases.

7) <u>High temporal, energetic and spatial stability</u> : required for reliable high accuracy experiments.

8) <u>Temporal coherence</u> : important for interrogating coherent optical effects.

II. SYNCHROTRON RADIATION SOURCES

An electron is an accelerating field loses energy by radiation. In a synchrotron or a storage ring, the electrons undergo radial acceleration by magnetic constraint. A low velocity electron will radiate as a normal Hertzian electric dipole (Fig. 2a) but when the electron velocity is high, a severe distortion of the radiation pattern occurs, due to relativistic effects. The zeros of the radiation pattern then occur at angles $\Theta = (1-v^2/c^2)^{1/2}$ from the direction of motion. The radiation pattern as seen by a stationary observer is indicated in Fig. 2b. Because of the relativistic

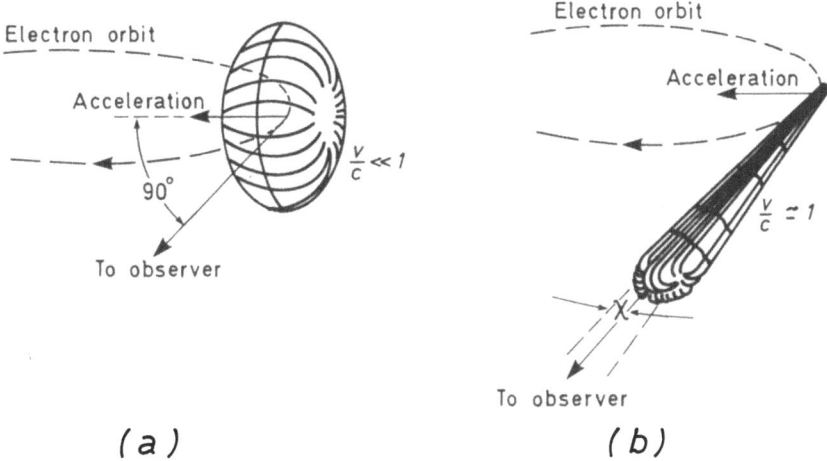

(a)   *(b)*

Fig. 2. Schematic radiation patterns for electrons in circular orbit (a) at low velocities ; b) a velocities approaching that of light.

transformation, the power radiated, which increases as E^4/R^2 (E = electron energy, R = radius of curvature of the electron orbit) is projected almost completely into a very small forward cone of the order of a milliradian in angle. The emitted power spectrum comprises the Fourier components of the synchrotron radiation pulse. The fundamental frequency is Doppler shifted from the MHz region into higher frequencies and the harmonics blur into a continuum, so that the observed spectrum extends from the infra-red up to the U.V. and soft X-ray regions. The divergence of the resulting light is comparable to that of a laser.

One must distinguish between electron synchrotron and electron storage rings as synchrotron sources. An electron synchrotron is modulated by three superimposed frequencies, the electron injection frequency (typical value $\sim$ 50 Hz), the electron orbital frequency ($\sim$ 1 MHz) and the cavity radiofrequency used to accelerate and restore energy lost by radiation ($\sim$ 400 Hz). The electrons are injected periodically ($\sim$ every 20 ms), and are accelerated with a periodically varying magnetic field. There is consequently a continuous time variation in the intensity, frequency and angular distribution of the synchrotron radiation. In an electron storage ring, an electron once injected is maintained in a stable orbit by a constant magnetic field. The only frequency of importance is the orbital frequency ν; the R.F. field usually operates at frequency ν or 2ν.

The basic relations between instrumental parameters and synchrotron radiation characteristics are as follows[4,5] (where the electron energy E is in GeV, the magnetic radius of curvature R is in meters, the electron current I in amps and the magnetic field B in kilo Gauss) :

(i) total power radiated by a relativistic electron

$$P = \int I(\lambda,\psi)\ d\lambda d\psi = 2/3\ \frac{e^2 c}{R^2}\ \frac{E^4}{(m_0 c^2)^4}$$

(ii) energy loss per revolution, per electron

$$\delta E(keV) = 88.5\ E^4/R$$

(iii) total power radiated

$$P_{tot}(kW) = 2.66\ E^3.B.I$$

(iv) critical wavelength

$$\lambda_c(\overset{\circ}{A}) = 5.59\ R/E^3 = 186.4/BE^2$$

(v) characteristic energy

$$\varepsilon_c(eV) = 2218\ E^3/R = 2.96 \times 10^{-7}\ \gamma^3/R$$
$$= 0.0665\ BE^2$$

(vi) mean emission angle $\Theta \approx \gamma^{-1}$

where $\gamma = E/m_0 c^2 = 1957\ E$

Fig. 3 gives the radiation spectrum of an electron moving in a curved trajectory, per GeV.

Characteristics of present-day and projected synchrotrons and storage rings can be found listed elsewhere[6,7].

Synchrotron radiation (SR) sources provide a unique combination of characteristics that are extremely useful as excitation sources for studies of molecular dynamics[8]. The following properties of current SR sources make them quasi-ideal :

1) tunability over the entire relevant energy range ;

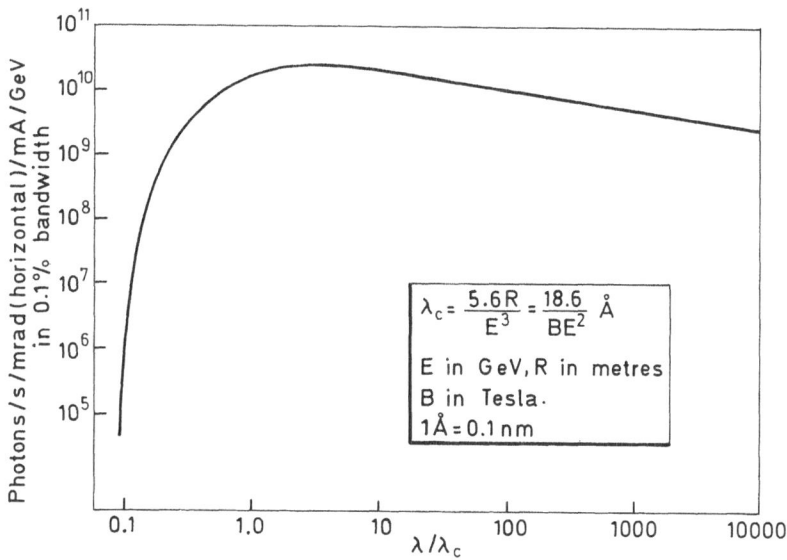

Fig. 3. Normalized radiation spectrum of an electron moving in a
curved trajectory per GeV.

2) resolution of the temporal pulse structure giving a pulse width of the order of 20-100 ps ; note that electron storage rings provide more useful temporal structure than electron synchrotrons. In storage rings the electron bunch length is typically 50 ps - 1 ns and the time between bunches can range from a few ns to a few μs.

3) energy resolution 0.05 $\overset{\circ}{A}$ or better in the range 5-100 eV ;

4) intensity (brightness) of the source with available fluxes of about 10^{11} photons/0.1 eV sec at the exit slit of the monochromator, impinging on the sample ;

5) high degree of spatial collimation ($\sim$ 1 mrad) ;

6) complete linear polarization ;

7) temporal and spatial stability

Future SR sources will be characterized by higher beam intensity and will generate quasi-monochromatic SR radiation, as discussed later in this volume[10], while more sophisticated detection methods[2] should result in time-resolved information on the ps time scale.

Two basic disadvantages of SR sources in the present context should be noted :

(i) the intensity of present SR sources is adequate for one-photon selective excitation but is too weak for multiphoton excitation. However, this limitation can be used to some advantage, as compared with intense lasers, since molecular dynamic studies resulting from SR excitation will not suffer from the additional complications which often result from multiphoton processes.

(ii) studies of coherent optical effects require temporal coherence of the radiation which cannot be achieved with current SR sources. Although the use of wigglers in currently planned SR sources will result in quasi-monochromatic radiation[10,11], it is still an open question whether the temporal coherence properties[12] of future SR radiation sources will be adequate for coherent optical effects to be interrogated.

III. SYNCHROTRON, LASER AND OTHER VUV SOURCES

The complementary nature of SR and laser sources is most marked in the high energy region above 5 eV and we will limit further discussion to this region.

Three types of alternative excitation sources must be considered for the high energy range above 5 eV. They are vacuum ultra-violet (VUV) sources[13-15], laser generated plasmas[16] and VUV lasers[17-20].

A comparison between the characteristics of SR sources and VUV discharge and plasma sources[1] shows that the discharge and plasma sources have the following disadvantages : a) relatively narrow tunability range except for the BRV source which can be used between 4 and 250 eV[15] ; b) long pulse length - the shortest pulse length obtained from a discharge source is 200 ns from the BRV plasma source[15], which is not very adequate for the study of time-resolved dynamics ; c) low degree of spatial collimation ; d) lack of polarization characteristics. Discharge and plasma sources also suffer from two of the shortcomings of SR sources i.e. moderate intensity and lack of temporal coherence.

We now compare SR with VUV lasers as potential photoselective excitation sources. Vacuum ultraviolet lasers available in 1980 have been listed, and their properties discussed, by Wallace[17]. Some newer developments have occurred since that date[18-22]. Table I presents a comparison between those characteristics of some representative VUV lasers and SR excitation sources which are relevant for intramolecular studies. Various techniques are used for generation of the VUV radiation : high-power gas discharge and electron-beam pumped lasers (e.g. ArF and F_2), and non-linear optical (harmonic) generation and frequency mixing from visible and ultraviolet dye lasers. The latter techniques provide the possibility of spectral tuning by the fairly broad frequency scan available with dye lasers. It should be stressed that the complete spectral region between 970 and 2100 Å can now be spanned by frequency mixing and harmonic generation techniques[17,22]. VUV laser operation has also been achieved at $\lambda < 970$ Å using harmonic generation techniques[17] but the low intensity of such lasers make them of limited use for the purpose of molecular dynamics studies.

Comparison between SR and VUV laser sources shows the following :

(i) SR sources will undoubtedly be superior to VUV laser sources with respect to tunability over the broad energy range required for studies of molecular dynamics.

(ii) the time resolution of SR sources is superior to that of current VUV lasers but some future VUV lasers will accomplish sub-picosecond time resolution.

(iii) the effective energy resolution of VUV lasers is better than that of SR sources.

Table 1. Characteristics of representative VUV lasers and synchrotron radiation sources.

Photon Source	Energy Region	Energy Resolution	Intensity (photons/pulse) [a]	Time Resolution (pulse length)
S.R. (present)[1]	10 Å – 100 μ [b]	0.02 Å (300–3000 Å) 0.15 Å (100–300 Å)	10^9 – 10^{10} [c] photons/Å sec	≥ 40 ps
S.R. (future ; with wigglers, undulators and Toroidal grating monochromators[1]	10 Å – 100 μ [b]	quasi-monochromatic pulses [d]	10^{12} – 10^{14} [c] photons/Å sec	∿ 100 ps [e]
ArF[23]	1932 Å	10 cm^{-1}	10^{18} – 10^{20}	55 ns
ArF with unstable resonator[24]	1932 Å	10 cm^{-1}	10^{15}	5 – 10 ns
F_2 [25]	1580 Å	< 25 cm^{-1}	10^{16}	15 ns
Mg vapour VUV harmonic generation 4-wave mixing[26,27]	1360 – 1600 Å 1200 – 1300 Å	0.1 cm^{-1} 0.1 cm^{-1}	10^{12} ≥ 10^8	10 ns
VUV harmonic generation (Mg vapour), with mode-locked dye laser (future)[1,17]	1360 – 1600 Å	0.01 cm^{-1}	10^{14}	0.3 ps
VUV harmonic generation Ar(f)[21]	970 – 1025 Å	1.7 cm^{-1}	10^{10}	30 ns
4-wave difference frequency mixing in Xe gas[18]	1520 – 2000 Å	≤ 0.1 cm^{-1}	4 x 10^{16}	4 ns
4-wave mixing in Hg[20]	1170 – 1220 Å	0.04 cm^{-1}	10^9 – 10^{12}	5 ns

[a] Intensity in photons/pulse unless otherwise stated.

[b] Energy region required for studies of molecular dynamics. Appropriate synchrotron sources can emit significantly at λ < 10 Å and
λ > 100 μ.

[c] Intensity at monochromator exit in 10–3000 Å spectral region.

[d] Photon energy band depends on wiggler/undulator characteristics.

[e] The pulse length is not expected to be shortened much below 40 ps.
A possible 1 ps time resolution may be obtained by using sophisticated detection methods[1,2].

[f] A pulsed Argon jet is used as frequency tripling medium of radiation in the 3070–2920 Å region produced by a frequency doubled
Nd : YAG-pumped dye laser[21].

(iv) the intensity of current VUV lasers is greater than current SR sources, making possible multiphoton excitation by VUV lasers.

(v) the polarization characteristics of both SR and VUV laser sources are excellent for molecular dynamics studies.

(vi) the temporal coherence of VUV lasers permits interrogation of coherent optical effects in the high energy range which cannot be carried out using SR.

This analysis shows that VUV lasers will be extremely useful for some photoselective studies carried out over a very narrow spectral region in the 5-13 eV energy range for one-photon excitation and 15-30 eV for multiphoton excitation. These studies include :

- energy-resolved experiments under ultra-high spectral resolution.

- time-resolved experiments with subpicosecond time resolution using ultra-short pulses from future VUV lasers.

- studies of coherent optical effects.

- studies of multiphoton excitation in the 15-30 eV range.

On the other hand, the inherent limitations of VUV lasers, as compared to SR sources are as follows :

- the output of VUV lasers consists of a single spectral line or is limited to tuning over a narrow energy range.

- the available energy range for VUV lasers is relatively small i.e. 5-13 eV for one photon and 10-30 eV for multiphoton excitation.

The use of VUV lasers in the area of molecular dynamics cannot be considered as an alternative to SR excitation. Rather, the utilization of VUV laser sources will provide useful information concerning one-photon and multiphoton excitation, ultrashort decay times and coherent effects, all interrogated over a narrow energy range, or at various selected energies. This information will be supplementary and complementary to that obtained using SR radiation over a broad energy range.

REFERENCES

1. J. Jortner and S. Leach, J. Chim. Phys. 77, 7 (1980).
2. I.H. Munro and A.P. Sabersky in "Synchrotron Radiation Research" (ed. H. Winick and S. Doniach), Plenum Press N.Y. (1980), p. 323.
3. C.V. Shank, R.L. Fork, R. Yen, R.H. Stolen, W.J. Tomlinson,

Appl. Phys. Lett. $\underline{40}$, 9 (1982) ; C.V. Shank, Science $\underline{219}$, 1027 (1983).

4. D. Ivanenko and I. Pomeranchuk, Phys. Rev. $\underline{65}$, 343 (1944).

5. J. Schwinger, Phys. Rev. $\underline{70}$, 798 (1946) ; $\underline{75}$, 1912 (1949).

6. E.E. Koch, J. Chim. Phys. $\underline{77}$, 21 (1980).

7. H. Winick in "Synchrotron Radiation Research" (ed. H. Winick and S. Doniach) Plenum, N.Y. 1980, p. 27.

8. "Perspectives of Synchrotron Radiation Applications to Molecular Dynamics and Photochemistry", Workshop proceedings edited by J. Jortner and S. Leach, J. Chim. Phys. $\underline{77}$, 1-57 (1980).

9. P. Dagneaux, C. Depautex, P. Dhez, J. Durup, Y. Farge, R. Fourme, P.M. Guyon, P. Jaeglé, S. Leach, R. Lopez-Delgado, G. Morel, R. Pinchaux, C. Vermeil and F. Wuilleumier, Ann. de Physique 14e série, $\underline{9}$, 9 (1975).

10. S. Leach, this volume.

11. J.E. Spencer and H. Winick in "Synchrotron Radiation Research" (ed. H. Winick and S. Doniach), Plenum Press, N.Y. (1980), p. 663.

12. C. Benard and M. Rousseau, J. Opt. Soc. Am. $\underline{64}$, 1433 (1974).

13. J.A.R. Samson "Techniques of Vacuum Ultraviolet Spectroscopy", Wiley, N.Y. (1967).

14. A.N. Zaidel' and E.Ya. Shreider "Vacuum Ultraviolet Spectroscopy", Ann Arbor-Humphrey Sci. Publ., Ann Arbor, Michigan (1970).

15. G. Ballofet, J. Romand and B. Vodar, C.R. Acad. Sci. (Paris) $\underline{252}$, 4139 (1961) ;
    H. Damany, J.Y. Roncin and N. Damany-Astoin, Appl. Opt. $\underline{5}$, 297 (1966) ;
    J.N. Fox and J.E.G. Wheaton, J. Phys. $\underline{E6}$, 655 (1973) ;
    E. Boursey and H. Damany, Appl. Opt. $\underline{13}$, 589 (1974) ;
    T.B. Lucatorto, T.J. McIlrath and G. Mehlman, Appl. Opt. $\underline{18}$, 2916 (1979).

16. P.K. Carroll, E.T. Kennedy and G. O'Sullivan, Opt. Lett. $\underline{2}$, 72 (1978) ;
    C.G. Mahajan, E.A.M. Baker and D.D. Burgess, Opt. Lett. $\underline{4}$, 283 (1979) ;
    P.K. Carroll, E.T. Kennedy and G. O'Sullivan, Appl. Opt. $\underline{19}$, 1454 (1980).

17. S.C. Wallace, Adv. Chem. Phys. $\underline{47}$, 153 (1981).

18. J. Hager and S.C. Wallace, Chem. Phys. Lett. $\underline{90}$, 472 (1982).

19. J. Bokor, T. Zavelovich and C.K. Rhodes, Phys. Rev. $\underline{A21}$, 1453 (1980) ;
    J. Bokor, R.R. Freeman, R.L. Panock and J.C. White, Opt. Lett. $\underline{6}$, 182 (1981) ;
    H. Junginger, H. Puell, H. Schingraber and C.R. Vidal, I.E.E.E. J. Quantum Electron. $\underline{17}$, 557 (1981).

20. R. Mahon and F.S. Tomkins, I.E.E.E. J. Quantum Electron., $\underline{QE-18}$, 913 (1982) ;
    F.S. Tomkins and R. Mahon, Opt. Lett. $\underline{7}$, 304 (1982).

21. E.E. Marinero, C.T. Rettner, R.N. Zare and A.H. Kung, Chem.
    Phys. Lett. (1983), in press.
22. K.H. Welge, this volume.
23. J.M. Hoffman, A.K. Hayes and G.G. Tisone, Appl. Phys. Lett. $\underline{28}$,
    538 (1976).
24. T.M. McKee, B.P. Stoicheff and S.C. Wallace, Appl. Phys. Lett.
    $\underline{30}$, 278 (1977).
25. H. Pummer, K. Hohla, M. Diegelmann and J.P. Reilly, Optics
    Comm. $\underline{28}$, 104 (1979).
26. S.C. Wallace and G. Zdasiuk, Appl. Phys. Lett. $\underline{28}$, 449 (1976) ;
    A.C. Provorov, B.P. Stoicheff and S.C. Wallace, J. Chem. Phys.
    $\underline{67}$, 5393 (1977).
27. T.J. McKee, B.P. Stoicheff and S.C. Wallace, Optics Lett. $\underline{3}$,
    207 (1978).

ENHANCING SYNCHROTRON RADIATION : WIGGLERS, UNDULATORS AND THE

FREE ELECTRON LASER

Sydney Leach

Laboratoire de Photophysique Moléculaire du C.N.R.S.
Bâtiment 213
Université Paris-Sud - 91405 Orsay Cedex, France

ABSTRACT

   Wigglers, undulators and free electron laser devices are presented and their characteristics discussed. Technological developments in progress will make these devices extremely useful and versatile high intensity, tunable photon sources of synchrotron radiation.

I. INTRODUCTION

   Properties of synchrotron radiation as obtained from normal
electron synchrotron and storage ring sources have been presented
elsewhere in this volume[1]. We are here concerned with three devices
which serve to enhance the use of synchrotron radiation : wigglers,
undulators and the free-electron laser.

II. WIGGLERS

   The object of wiggler systems is to obtain synchrotron radiation that is of much greater intensity and of different spectral
distribution than can be obtained with normal bending magnets in
a storage ring or synchrotron.

   A wiggler is a magnet device which can be inserted in a
straight section of an electron storage ring or synchrotron and
which causes the electron beam to execute a trajectory whose local
radius of curvature is smaller than in the normal ring bending
magnets. The wiggler field produces tranverse acceleration with no
overall deflection or displacement. With a magnetic field higher

than that of the ring magnets, the result is to increase the synchrotron radiation critical energy by a considerable factor and to
shift the overall spectrum to a higher energy by the same factor.
In fact one could also use a lower magnetic field on the wiggler
and so produce a "softer" synchrotron radiation spectrum without
altering the electron energy. If the wiggler is designed so that
the oscillation of the electron beam is comparable to the transverse beam dimensions, the radiation from all poles are then superimposed and the photon intensity is greatly enhanced.

Fields in the ring magnets are generally smaller than 12 kG
but wiggler fields of 20 kG or more are possible using iron-core
magnets and 50 kG or more with superconducting magnets[2]. Samarium-
cobalt permanent magnets are now considered to be among the best
materials for wigglers and undulators[3]. Wigglers have been constructed or are being planned at most storage rings and at a few synchrotrons throughout the world[2].

III. UNDULATORS

The emission spectrum of standard wigglers producing one or
a small number of transverse oscillations of the electron beam is
reasonably continuous in spectral distribution. However, multipe-
riodic wigglers (e.g. 30-100 periods) can produce a synchrotron
radiation spectrum which consists of one or a few narrow bands
emitted at a given angle. This radiation results from interference
between the electromagnetic fields emitted by the same electron at
different points on its trajectory ; contributions from different
electrons are incoherent. These interference wiggler devices are
known as undulators[2].

Two undulator configurations are usually considered : (i) a
transverse sinusoidal magnetic field with period λ_0 and maximum
field amplitude B_0, and (ii) a helical magnetic field in which the
field amplitude remains constant but the direction of the field
vector rotates about the axis of the undulator as a function of the
distance along the axis. The fundamental wavelength emitted in the
case of a planar undulator for electrons travelling parallel to its
axis is given by[4-6] :

$$\lambda_f = \frac{\lambda_0}{2\gamma^2} (1 + \frac{K^2}{2} + \gamma^2\theta^2)$$

where λ_0 = undulator period in mm, θ = angle of observation with
respect to the mean electron trajectory and the deflection parame-
ter $K = \alpha\gamma = 0.934\ B_0\lambda_0$ where α is the maximum deflection angle and
B_0 (kG) is the maximum magnetic field on the trajectory.

The harmonic content increases with the magnetic field ampli-

tude. The form of the spectrum depends on K, which is independent
of the electron energy[7]. For K << 1 only one spectral harmonic is
emitted ($\lambda = \lambda_0/2\gamma^2$) (Fig. 1) ; its wavelength is independent of B_0
and the power emitted is proportional to B_0^2 ; the wavelength spread
is small if the radiation is observed through a pinhole. When K $\gtrsim$ 1
the spectrum includes many harmonics ; under certain conditions,
their envelope ressembles a normal synchrotron radiation spectrum.
If the magnetic field configuration is helical instead of transverse, the dependence on θ and K are very similar except that at $\theta = 0$
only the fundamental is observed   and polarization is circular[7] ;
at $\theta > 0$ harmonics with elliptical polarization are emitted, the
polarization becoming linear for $\theta = \gamma^{-1}$.

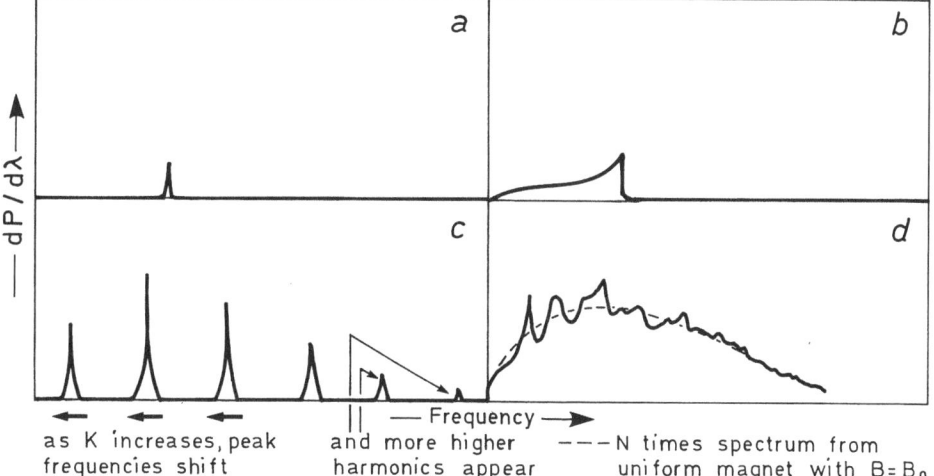

Fig. 1.   Schematic representation of spectral characteristics of
          undulators.
          a) K << 1, Σ << 1 : weak field, well collimated beam
          b) K << 1, Σ $\gtrsim$ 1 : weak field, σ' large or no pinhole
          c) K $\gtrsim$ 1, Σ << 1 : strong field, well collimated beam
          d) K $\gtrsim$ 1, Σ $\gtrsim$ 1 : strong field, σ' large or no pinhole

          K = Deflection parameter (see text)

          $\Sigma = \left[(\gamma\sigma')^2 + \gamma^2 - \frac{\Omega}{\pi} \right]^{1/2}$ is a parameter which takes

          into account the angular spread of the electrons (σ')
          and the acceptance angle of the detector (Ω).

The gain in power emitted in a solid angle $2\gamma^{-2}$ x $2\gamma^{-2}$, as compared with normal bending magnets, is of the order of N, the number of undulations. An increase of $\approx N^2$ in spectral brilliance can be obtained for a well collimated beam. The spectral width is of the order of N^{-1} for a small beam aperture and decreases with increasing θ. Undulators with values of N up to about 100 have been built[2,3,8] or are being planned[2].

IV. FREE ELECTRON LASER

Theoretical analysis of free-free transitions of electrons moving in undulating magnetic fields has much developed since the original work of Ginzburg[9] and Motz[2]. Its renewed importance is related to the development of Free-Electron Lasers (FEL) using free-free transitions of relativistic electrons. The present theoretical situation has been recently summarized by Fedorov[10]. Two main types of FEL have been proposed : (i) the low electron density FEL in which electrons individually experience the undulator magnetic field and the photon field, and (ii) the high electron density FEL where electrons interact collectively with the undulator and photon fields. Dense electron beams, for which resonance effects appear at plasma frequencies[11,12], cannot be achieved in storage rings. (A FEL using low energy (1.2 MeV) high current (25 kA) beams has been operated giving emission in the far I.R. at 400 μ[12]). Our discussion will therefore only consider the low density electron beam case, using planar field undulators.

In the electron reference frame, the periodic undulator field appears as an electromagnetic field corresponding to virtual photons. It can thus amplify a real photon field travelling in the same direction as the relativistic electrons, the two fields being coupled by the electron charges (stimulated Compton scattering[13]). The quantum mechanical description by Madey[14] led to his building the first FEL (in the Stanford linear accelerator)[15].

The gain in a FEL is inversely proportional to the square of the linewidth of spontaneous radiation. The latter has two components (i) a homogeneous width due to the finite interaction time of electrons in the undulator, (ii) an inhomogeneous width due to the angular divergence and velocity spread of the electrons as well as to the variation with position of the strength of the periodic magnetic field. The maximum gain G_{max} is proportional to $\rho_e B_0^2 \lambda_0^{3/2}$ $\lambda^{3/2} N^2 L$ where $L = N\lambda_0$ is the undulator length and ρ_e is the electron density (assuming that the electron energy distribution is Gaussian).

Gain measurements were first made by Madey and his coworkers at Stanford[16]. They used a superconducting transverse helical undulator, period λ_0 = 3.2 cm, and electrons in a linear accelerator.

Spontaneous gain was observed for 10.6 μ CO_2 laser radiation with
24 MeV electron energy. A gain of 7 % per pass was measured with an
electron current of 70 mA. A 12 m long optical cavity was then added
to the undulator so as to make a FEL ; with 43.5 MeV electrons laser
emission was observed at 3.417 μ with an average power of 0.36 W,
peak power 7 KW[15].

Gain measurements on visible radiation have been made at Orsay
using a 23 period superconducting undulator (4 cm wavelength and
4 kG design field) inserted in a straight section of the ACO storage ring, and an Ar^+ laser. Spontaneous emission was measured for
electron energies of 240 MeV and 150 MeV[17-19]. Fig. 2 shows the
gain and spontaneous emission curves as a function of electron
energy at 4880 Å. The negative part of the gain curve corresponds
to net absorption ; zero gain coincides with the spontaneous emission maximum at the resonance energy. The curves behave as expected
from theory which predicts that the gain will be proportional to
the energy derivative of the spontaneous emission[20]. The largest
measured peak gain, averaged over the laser mode was $\hat{G}$ = 4.3 x 10^{-4}
per pass. In order to convert the set-up to a FEL (Fig. 3), cavity
mirrors of exceptional quality are required in the visible and
ultraviolet regions . The technical problems of producing adequate,
radiation resistant, mirrors have not yet been solved. Higher gains
are expected through use of an optical klystron[21] instead of a normal undulator. This consists of two identical undulators between
which is a space in which a magnetic field induces a single large
wiggle in the electron trajectory. Optical klystrons are being tested in Orsay[22].

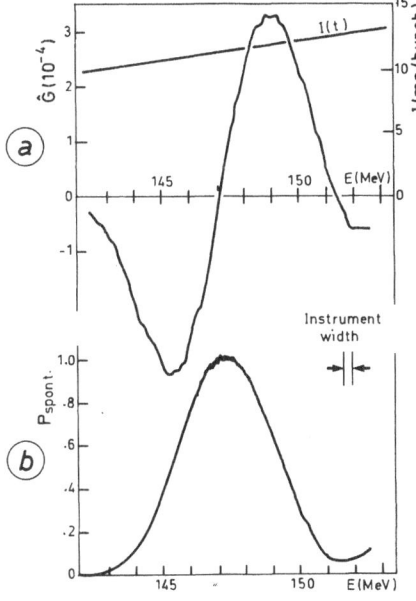

Fig. 2. a) Grain and b) Spontaneous
spectrum as a function of
electron energy. I(t) indicates the fall-off of SR
intensity during the ex-
periment as the electron
is lowered.

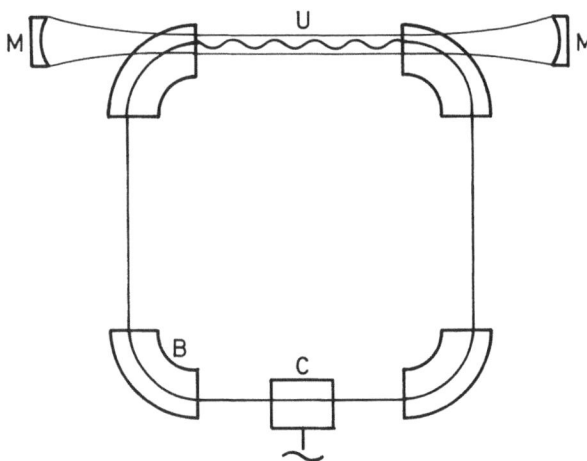

Fig. 3. Schematic set-up of a storage ring FEL.
         U = Undulator ; M = Optical cavity mirror ; B = Bending
         magnets and focusing elements ; C = Radiofrequency cavity.

     The FEL beam has spatial and time structures that reflect the
electron beam structure. The radiation pulse length is proportional
to the electron bunch length ℓ_e, and the radiation frequency width
is given by $\delta\omega = 2 \pi c/\ell_e$. Individual laser modes have linewidths
which are approximately equal to the inverse of the correlation
time. The practical limit to the correlation time, and therefore
to the widths of individual lines in the spectrum, is set by mirror
microphonics. The laser power $P_L = G_{max} \times P_{RF}$ where P_{RF} is the power
given to the electron beam by the RF cavity.

     The properties of FEL which are considered to be possible with
existing or foreseen improved technology can be summarized as follows[23]. The operating wavelength is a continuous function of the
electron energy and the magnetic field in the undulator ; the range
is from the mm region to the vacuum ultraviolet, with $\lambda = 500 \overset{\circ}{A}$
being a lower wavelength limit beyond which gain could never be
sufficient. It should be possible to tune the emitted radiation
over about one octave in the I.R., visible and U.V. regions. The
best electron source will vary with wavelength region : storage
rings in the visible and U.V., microtrons in the I.R., and Van de
Graaf machines in the mm region. Average powers of the order of
1 KW or more are expected and much higher values (MW) are contem-
plated with improved technology ; overall net energy conversion
efficiencies of the order of 1 % are expected and could possibly
be an order of magnitude greater. In the case of a FEL operated
with an electron beam from an RF accelerator the optical cavity
length must be set to a multiple of the electron bunch spacing ;

the optical field in such a FEL is mode-locked by the periodic variation in electron current ; picosecond pulses can be obtained at
high repetition rates (10 MHz - 1GHz). All the usual laser techniques
used to minimize linewidths can be applied to the FEL so that it
should be possible to obtain individual lines with $\Delta\nu \lesssim$ 1 MHz. An
important advantage of the FEL is the possibility for it to operate
simultaneously at different frequencies either with a single undulator, for closely related frequencies, or with a set of independent
undulators. Furthermore, a storage ring FEL also generates significant amounts of broadband incoherent synchrotron radiation in addition to the emitted (laser) coherent radiation ; this SR could be
of interest for independent experiments or for use as a secondary
source in conjunction with the laser.

Finally, we summarize some possible uses for free electron
lasers. 1) Far infrared : interesting wavelength range ; useful for
saturation pumping, multiphonon processes, surface state spectroscopy, solid state spectroscopy and photophysics. 2) U.V. - V.U.V. :
advantage can be taken of flexibility, power density and wavelength
range for studies on spectroscopy, multiphoton ionization, high
energy-vibronic states, low density targets such as transients e.g.
radicals, ions, muonium, positronium. 3) Pump and probe experiments could be carried out by FEL + S.R., FEL + separate laser, or
FEL operating in multifrequency mode. 4) Study of laser assisted
collisions and general photochemistry, in particular state and/or
species selective photochemistry using the multifrequency resources.
5) Industrial applications of photophysical and photochemical processes.

We end with a note of caution in stressing (i) that the output
of theoretical work on free-electron lasers far exceeds the experimental activity ; (ii) that much technological improvement is
required before the FEL properties discussed above become effective.
Progress will probably be most rapid in the mm and I.R. radiation
regions. Feasibility studies should be complete in many FEL areas
within the next three years and the direction of the field and its
real possibilities will then be more clear.

REFERENCES

1. S. Leach, this volume.
2. J.E. Spencer and H. Winick in "Synchrotron Radiation Research"
   (ed. H. Winick and S. Doniach), Plenum, N.Y. (1980) p. 663.
3. J.M. Ortega, C. Bazin, D.A.G. Deacon, C. Depautex and P. Elleaume,
   Nucl. Instr. and Methods (1982) in press.
4. H. Motz, J. Appl. Phys. 22, 527 (1951).

5. D.F. Alferov, Yu A. Bashmakov and E.G. Bessonov, Sov. Phys. Tech. Phys. 18, 1336 (1974).

6. A. Hofmann, Nucl. Inst. and Meth. 152, 17 (1978).

7. European Synchrotron Radiation Facility, Supplement II : The machine, (ed. D.J. Thomson and M.W. Poole), ESF, Strasbourg (1979) p. 52 et seq.

8. C. Bazin, Y. Farge, M. Lemonnier, J. Pérot and Y. Petroff, Nucl. Instrum. Methods 172, 61 (1980).

9. V.L. Ginzburg, Izv. Akad. Nauk (USSR), ser. phys. 11, 165 (1947).

10. M.V. Fedorov, Prog. Quant. Electr. 7, 73 (1981).

11. T. Kwan, J.M. Dawson and A.T. Lin, Phys. Fluids 20, 581 (1977).

12. D.B. McDermott, T.C. Marshall, S.P. Schlesinger, R.K. Parker and V.L. Granatstein, Phys. Rev. Lett. 41, 1368 (1978).

13. P.L. Kapitza and P.A.M. Dirac, Proc. Cambridge Phil. Soc. 29, 297 (1933).

14. J.M.J. Madey, J. Appl. Phys. 42, 1906 (1971).

15. D.A.G. Deacon, L.R. Elias, J.M.J. Madey, G.J. Ramian, H.A. Schwettman and T.I. Smith, Phys. Rev. Lett. 38, 892 (1977).

16. L.R. Elias, W.M. Fairbank, J.M.J. Madey, M.A. Schwettman and T.I. Smith, Phys. Rev. Lett. 36, 717 (1976).

17. C. Bazin, M. Billardon, D. Deacon, Y. Farge, J.M. Ortega, J. Pérot, Y. Petroff and M. Velghe, J. Physique Lett. 41, L547 (1980).

18. D.A.G. Deacon, J.M.J. Madey, K.E. Robinson, C. Bazin, M. Billardon, P. Elleaume, Y. Farge, J.M. Ortega, Y. Petroff and M. Velghe, I.E.E.E. Trans. Nucl. Sci. NS-28-3142 (1981).

19. D.A.G. Deacon, K.E. Robinson, J.M.J. Madey, C. Bazin, M. Billardon, P. Elleaume, Y. Farge, J.M. Ortega, Y. Petroff and M. Velghe, Opt. Comm. 40, 373 (1982).

20. J.M.J. Madey, Il Nuovo Cim. 50B, 64 (1979).

21. N.A. Vinokurov and A.N. Skrinsky, Proc. 10th Int. Conf. High Energy Charged Particle Accelerators, Serpukhov Vol. 2, 454 (1977).

22. M. Velghe, M. Bergher, C. Bazin, M. Billardon, D.A.G. Deacon, P. Elleaume, J.M.J. Madey, J.M. Ortega, Y. Petroff and K.E. Robinson, Prof. Int. Conf. Lasers'82, New Orleans, December 1982, in press.

23. J.M.J. Madey and J.N. Eckstein, Report of Workshop (Riva del Garda, 1979) on The Possible Impact of Free Electron Lasers on Spectroscopy and Chemistry (ed. G. Scoles).

PART II

LASER APPLICATIONS TO ANALYTICAL CHEMISTRY

ANALYTICAL CHEMISTRY METHODS BASED ON ABSORPTION OF LASER LIGHT

John C. Wright

Department of Chemistry
University of Wisconsin
Madison, Wisconsin 53706

INTRODUCTION

This part of the course is intended as a tutorial overview of the research that is currently going on in applying lasers to analytical chemistry. Since the field is large, many topic areas will of necessity be left out. More extensive reviews are available elsewhere (1, 2).

Analytical chemistry is basically the science of chemical measurements. The classical measurement for an analyst is the determination of the composition of a substance both qualitatively and quantitatively. This definition is quite restrictive because many other chemical measurements are performed, especially in modern-day research. Thus, the measurement of vibrational relaxation times can be quite important in characterizing a material. I will nevertheless concentrate the review on the more classical methods of analysis because other chapters will provide examples of the other measurements.

PRESENT METHODS OF ANALYSIS

To obtain an appreciation for laser methods, we must first understand the methods that are presently used. There are a very large number of methods that use almost every property of molecules or atoms that might be measured (3). I will describe only the most commonly used and widely applicable ones.

One might divide analyses into two categories: elemental and molecular. Elemental methods require decomposition of a substance to the atoms themselves and the subsequent analysis. An inductively

coupled plasma (ICP) provides a convenient approach (4). The sample
is dissolved and the solution is aspirated into a high temperature
plasma that has been formed by flowing argon gas through an intense
rf field generated by surrounding coils. The high temperatures cause
desolvation, atomization, and excitation of the sample. Measurement
of the atomic line positions and intensities provide information
about the identity and concentration of the elemental constituents.
Similarly, a high temperature flame can be used to atomize a sample
that is aspirated into it. The emission from the more easily excited
atoms can be used for qualitative and quantitative atomic analysis
(flame emission spectroscopy) or light from a hollow cathode lamp
made from the desired element can be passed through the flame to
measure the absorption of atoms in the flame (atomic absorption
spectroscopy) (3). The ICP has exceptionally low limits of detection
and the very sharp emission lines make the technique specific for
the atoms of interest. The very high temperatures make it quite
free from interference effects caused by other sample components.
It is more expensive to implement than flame emission or atomic
absorption, so both flame techniques find extensive usage for
chemical measurement.

   Measurement of uv or visible absorption is one of the most
common and widely applicable methods for molecular analysis (3).
Most molecules of interest either absorb or can be converted to a
molecule that absorbs. Absorption is sensitive enough for many
trace analysis problems but the wide absorption profiles of molecules in solution makes the technique subject to interference
effects from many other solution constituents. Chromatography provides a powerful tool for performing separations of sample constituents so that the subsequent detection can be non-specific for
particular molecules (3). If the sample can be volatilized at
reasonable temperatures (200-300°C), the chromatography is performed by flowing the sample in a carrier gas (the mobile phase)
over a stationary solid that has been coated with a wax or polymeric liquid (the stationary phase). If the sample cannot be
volatilized, the chromatography is performed by flowing the sample
dissolved in a solvent over a stationary phase whose surface has
been treated or coated to interact with the sample. Both gas
chromatography (GC) and liquid chromatography (LC) perform their
separation based upon the differences in interaction strengths between the sample constituents and stationary phase. A constituent
that interacts favorably with the stationary phase is swept along
more slowly than one that has little interaction so that different
constituents elute from a chromatographic column containing the
stationary phase at different times. Most molecular analysis problems can be solved with uv-visible absorption spectroscopy, GC,
and LC methods.

   If the sample is very complex and if the constituents of
interest are present in very low amounts, mass spectroscopy is

often used for detection and measurement (3). Molecules are ionized
by electron impact, chemical ionization, or field ionization,
accelerated through a potential difference, and passed through a
magnetic field to reach a detector. The amount of bending is related
to the mass/charge ratio of the ion. (Quadrupoles are also commonly
used to measure this ratio.) Mass spectroscopy is an expensive
method but it is very sensitive and gives a great deal of information about the identity of a molecule. In combination with GC, it
forms a very powerful, general, but expensive tool for molecular
analysis.

CONVENTIONAL ABSORPTION MEASUREMENT WITH LASERS

     Traditional blackbody sources perform well for uv-visible ab-
sorption spectroscopy and the laser will be hard-pressed to dis-
place them. Lasers do have some inherent advantages though. An absorption measurement requires comparison of the light intensity
passing through a blank with that passing through a sample. Fluctuations and noise in that comparison limit the method. Most measurements are usually limited by light source fluctuations but they are
ultimately limited by the shot noise of the source. The high spectral
powers of typical lasers make the shot noise contribution small.
A small 3 mW laser provides 10^{14} photons/sec so the shot noise contribution is 10^{-8} $Hz^{-1/2}$. Shot noise limited measurements have been
accomplished and absorbances of 10^{-8} have been measured (5-7). This
incredibly low value was reached by modulating an argon laser pumped
dye laser at 10 MHz. This high frequency is above the fluctuation
frequency of most noise sources and eliminates their contributions.
Wavelength modulation can be a very effective method for reducing
noise if the sample has sharp absorption features. Despite such
achievements, the traditional absorption measurement is limited to
absorbances of 10^{-2} to 10^{-4} by differences in surface reflections,
particulate scattering, or indices of refraction between the sample
and reference (2).

     In addition to reducing the noise in an absorption measurement,
one can increase the signal with a laser. The low angular divergence
of laser sources permits multiple passing through a sample cell
(8). Glass capillaries can also be filled with the sample so that
long path lengths can be achieved with reasonable amounts of
liquid (9-11).

OPTACOUSTIC SPECTROSCOPY

     Instead of measuring absorption by comparing two light intensi-
ties, the heating caused by absorption can be directly detected
(2). Heating effects form the basis for both optoacoustic and
thermal lensing spectroscopy. In optoacoustic spectroscopy, the
light source is modulated and passed through the sample cell (12,
13). Absorption results in heating which in turn causes the changes

in pressure that are detected as sound waves by a microphone. The
heat capacity and the thermal expansion coefficient of the sample
are important in determining the magnitude of the sound waves. The
signal levels scale with the power of the light source. Gas phase
samples are most easily measured. The sample is introduced into a
cell containing a microphone and heating produced in the gas is
transmitted directly to the microphone (12, 14, 15). Resonances
within the cell can be used to great advantage in improving the
signal levels (16). The modulation frequency must be tuned to one
of the cell resonances. These resonances will vary from sample to
sample because of differences in the gas characteristics. Typical
microphones produce 1 V/torr signals (17). Thus if 0.2 μV signals
can be detected, 0.2 μTorr pressure waves can also be detected.
Typically such pressure waves correspond to 10^{-9} J of absorbed
energy. If the source provides 10 mJ of energy, an absorbance of
10^{-7} is measurable. Such values are typical of optoacoustic gas
phase measurements. A value of 10^{-9} has actually been measured
(12).

Solid samples cannot be measured as easily by optoacoustic
methods (13). A solid sample is typically placed within a closed
chamber and an intermediary gas is required to transmit surface
heating of the solid to the microphone as pressure waves. Again
there is a strong dependence on the thermal properties of the solid.
The detection limits are much poorer for solids because scattering
from an opaque sample surface can be absorbed by the cell walls
creating a background that obscures low signals (40).

Liquids have been measured either by direct immersion of the
microphone or by acoustically coupling a sample cuvette to the
microphone through a quartz disk (19, 20). Absorptions of 10^{-7} can
be measured for liquid samples (19). This value is not as low as
gases but it is much better than solids.

The lowest levels in optoacoustic measurements are limited by
background absorption either by the cell windows or the solvent itself. Pulsed lasers with gated detection are especially useful for
discriminating against window absorption since sound takes longer
to reach the microphone from the cell windows (17). High laser intensity stability is needed if solvent absorption is a serious problem. If the solvent absorbs appreciably, an optoacoustic method
can give poorer results than conventional transmission measurements
(2).

In contrast with conventional transmission measurements, an
optoacoustic method is strongly dependent on the thermal properties
of the sample. Aqueous samples are particularly poor because of the
high heat capacity of water. Absorbances of 5 x 10^{-5} can be detected in water compared with 10^{-7} in favorable solvents (2). On
the other hand, reflection losses and differences in index of re-

fraction do not cause problems in optoacoustic spectroscopy. The
effects of particulate scattering are also reduced greatly. Again in
contrast to transmission measurements, increases in sample pathlength do not increase signal since the volume sampled by the microphone is limited. If optoacoustic methods are used in applications
requiring a flowing sample as an LC detector would, one must overcome the additional acoustical noise generated by the flow and
pumps.

THERMAL LENSING SPECTROSCOPY

A laser beam with a Gaussian intensity profile will cause un-
even heating effects across its profile as it passes through a
sample (21-31). Since a heated liquid or gas has a lower density
than an unheated one, the index of refraction will be lower. Thus
a negative lens will be produced in the sample by the Gaussian
laser profile which will affect the propagation of the beam. The
effect will be time dependent and will reach a steady state when
the heating of the laser and the cooling by thermal conduction to
the bulk of the sample equalize. Thus the thermal lensing will depend on the sample density (ρ), specific heat (C), the variation
of the index of refraction with temperature (dn/dT) and the thermal
conductivity of the sample (k). The time dependent focal length of
the thermal lens can be expressed as (29)

$$f(t) = f(\infty) \; \frac{1 + t_c}{2t} \qquad\qquad [1]$$

where
$$t_c = \frac{\rho \; Cw^2}{4k}$$

$$f(\infty) = \frac{\pi \; k \; w^2}{2.303 \; PA \; dn/dT}$$

w = beam size
P = laser power (watts)
A = sample absorbance.

In order to increase the thermal lensing effect, the laser is
usually focused within the sample. The influence of the thermal
lens on the laser beam then depends upon its position relative to
the focal spot of the laser. If for example the sample is placed
right at the laser focus, there is no effect on the beam. If the
sample is placed too far from the focus, the thermally lensing is
not as efficient. There is an optimum when the sample is placed one
confocal parameter on either side of the focus. The thermal lensing
effects are usually monitored by placing a pinhole in the center of
the beam and measuring the intensity through the pinhole (I_{bc}). At
the optimum position, one can express I_{bc} after the thermal lens
is established ($I_{bc}(\infty)$) to that before the thermal lens is created
($I_{bc}(o)$) as (31).

$$\frac{I_{bc}(\infty) - I_{bc}(o)}{I_{bc}(o)} = \frac{2.303 \ P \ \frac{dn}{dT} \ A}{\lambda \ k} = 2.303 \ EA \qquad [2]$$

where λ = laser wavelength

$$E = \frac{P \ \frac{dn}{dT}}{k} \ .$$

One can compare this equation with a conventional absorption measurement where absorbance is defined relative to the incoming light (I_o) and outgoing light (I) as $I/I_0 = 10^{-A}$. At small absorbance, one has $I_0 - I/I_0 = 2.303 \ A$. E in equation 2 represents an enhancement factor of thermal lensing over transmission measurements (21,30). Note that E can be larger or smaller than unity. It may seem strange that thermal lensing can give less effect than conventional absorption until one realizes that $I_{bc}(\infty)$ and $I_{bc}(o)$ represent intensities at two different times, not in two different places. Both $I_{bc}(\infty)$ and $I_{bc}(o)$ are attenuated by absorption that is occurring in the sample. Standardization is done more easily in thermal lensing measurements because the reference level is measured without removal of the sample. Typically, the laser is chopped and $I_{bc}(o)$ is measured immediately after the laser turns on and $I_{bc}(\infty)$ is measured after the thermal lens is established. This two point measurement procedure can be replaced by fitting the entire thermal lens transient to equation [1] (29). The larger number of points provides a factor of 30 improvement in the lowest measurable absorbance. Equation [2] also shows that the enhancement depends linearly on the laser power. Typical E/P values for different solvents are 8.93, 3.06, 0.21 mW^{-1} for CCl_4, methanol, and H_2O respectively (21). Again, water has poor thermal properties for thermal lensing spectroscopy (2).

A clever method can be used for correcting for solvent absorptions using a reference cell (31). In conventional absorption methods, the sample and reference measurement are done sequentially. In thermal lensing methods, the two can be done simultaneously. If a thermal lens is placed one confocal parameter before the focus of a laser beam, $I_{bc}(\infty)$ will be larger than $I_{bc}(o)$ because the diverging thermal lens will effectively move the focal spot back toward the pinhole and detector. Similarly, if the thermal lens is placed one confocal parameter beyond the focus, $I_{bc}(\infty)$ will be smaller than $I_{bc}(o)$ because the thermal lens adds divergence to an already diverging beam. If identical samples are placed one confocal parameter before and after the focus, their effect will cancel and no net change will be observed in $I_{bc}(\infty)$. Any differences in samples will be seen. Thus, one cell can be the sample and the other the reference so solvent absorption can be automatically compensated.

Thermal lensing has measured absorbances of 7×10^{-8} in a CCl_4 solvent with a 160 mW Ar^+ laser power (29). This measurement is a

favorable case and the enhancement factor was 868. Thermal lensing
is thus capable of measuring adsorbances as low as optoacoustic
methods are capable. It also has the same sharp dependence on the
sample's thermal properties. Reflection losses and differences in
index of refraction do not affect the measurement but particulate
scattering can cause problems. A flowing sample also causes problems
because of the distortions that occur in the thermal lens (2).

INTRACAVITY ABSORPTION SPECTROSCOPY

An intracavity absorption measurement relies again upon making
a transmission measurement (32-35). The effects of sample absorption
are accentuated though by three effects which become important when
the sample is placed inside the laser cavity (36-38). The multiple
passes that occur within a laser cavity increase the effective
sample pathlength. If the laser is operating close to threshold,
the laser output power becomes a sensitive function of the cavity
losses so any sample absorption is further amplified. Finally if
the sample has much sharper lines than the gain bandwidth of the
laser, the laser modes within the sample absorption profile will
experience larger losses and other modes will become more favored.
The same factors which increase the sensitivity to absorption also
make the laser output unstable so that quantitation of the measurement becomes difficult (2). Even small changes in the laser can
cause large changes in the output intensity. We shall concentrate
primarily on measurements where the sample absorption is broad
since that represents the typical situation in analytical measurements. In this case, the measurement is made by comparing the laser
output intensity when the sample is in the cavity with that of a
reference material. If the laser power in the cavity is different
between those two measurements because the sample and reference
are absorbing differently, the beam focusing and positions in the
cavity can change so the two measurements are not equivalent.
Thermal lensing effects can also contribute substantially to the
output changes (39). Intracavity absorption measurements therefore
do not necessarily have a simple relationship between output laser
power and sample absorption.

A successful solution to these problems is to use nulling
techniques to maintain constant intracavity powers (39, 40). An
electronically adjustable Pockel's cell is placed in the cavity of
an Ar^+ laser pumped c.w. dye laser along with the sample so the
intracavity losses can be continuously adjusted. Thermal lensing
effects are eliminated by chopping the laser and measuring the output before any thermal lenses are established. The laser is operated 10 - 30 % above threshold so that stable output powers are
obtainable and larger dynamic ranges are available for measurement. The losses that must be added to maintain constant power
are linearly related to the sample absorbance with this approach.
It has been possible to detect absorbances of 5 x 10^{-5} correspond-

ing to a 2 x 10^{-9} $\underline{M}$ Fe concentration (or 90 parts per trillion)
(39, 40). Since intracavity absorption is basically a transmission
measurement, the thermal properties of the sample are not important
and aqueous samples can be measured as easily as other solvents
(2). Samples can be flowed through the measurement cell without
difficulties. These measurements are sensitive to reflection losses,
changes in refractive index, and particulate scattering. Furthermore, they are expensive and difficult to implement but can be
attractive for specific applications.

CONCLUSIONS

    The laser has provided a number of new alternatives to ab-
sorption measurement. None of them is best for all situations.
Thermal lensing and optoacoustic methods generally work at the
lowest absorbances and they are simple and inexpensive. They are
strongly dependent on the sample properties and do not work as well
for aqueous samples. Thermal lensing methods require lasers with
good spatial mode stability and therefore are more limited in wavelength tunability. Intracavity absorption is expensive and difficult
but works equally well in all solvents. It is usually not competitive with absorbances measurable by thermal lensing and optoacoustic methods except for aqueous systems. All of the methods
have difficulty operating as detectors for LC applications either
because they are sensitive to flow or they are difficult to use.
All of the methods are limited in the selectivity for specific
compounds by the broad absorption features usually encountered.

REFERENCES

1. J.C. Wright, Applications of Lasers to Chemical Problems
        edited by T.R. Evans, John Wiley Interscience (New York,
        1982).
2. T.D. Harris, Anal. Chem. 54, 641A (1982).
3. H.H. Bauer, G.D. Christian, and J.E. O'Reilly, Instrumental
        Analysis, Allyn and Bacon (Boston, 1978).
4. V.A. Fassel, Science 202, 183 (1978).
5. A. Owyoung, Opt. Commun. 22, 323 (1977).
6. B.F. Levine and C.G. Bethea, Appl.Phys.Lett. 36, 245 (1980).
7. B.F. Levine and C.G. Bethea, IEEE J. Quantum Electron. QE-16,
        85 (1980).
8. T.G. Kyle and B.G. Shuster, Appl. Opt. 17, 2659 (1978).
9. J. Stone, Appl.Phys.Lett. 20, 239 (1972).
10. J. Stone, IEEE J. Quantum Electron. QE-8, 367 (1972).
11. J. Stone, Appl. Opt. 17, 2876 (1978).
12. C.K.N. Patel, Science 202, 157 (1978).
13. A. Rosencwaig, Anal.Chem. 47, 593A (1975).
14. K.P. Koch and W. Lahmann, Appl.Phys.Lett. 32, 289 (1978).
15. L.B. Kreuzer, Anal.Chem. 46, 237A (1974).

16. L.J. Thomas, M.J. Kelly, and N.M. Amer, Appl.Phys.Lett. 32, 736 (1978).

17. D.R. Siebert,G.A. West and J.J. Barrett, Appl.Opt. 19, 53 (1980).

18. J.F. McClelland and R.N. Kniseley, Appl.Opt.15, 2658 (1976).

19. C.K.N. Patel and A.C. Tam, Appl.Phys.Lett. 34, 467 (1979).

20. E. Voigtman,A. Jurgensen, and J. Winefordner, Anal.Chem. 53, 1442 (1981).

21. J.M. Harris and N.J. Dovichi, Anal.Chem. 52, 695A (1980).

22. R.L. Swofford, Lasers in Chemical Analysis, G.M. Hieftje, J.C. Travis, and F.E. Lytle, Eds., Humana Press (New Jersey, 1981).

23. J.P. Gordon, R.C.C. Leite, R.S. Moore, S.P.S. Porto, and J.R. Whinnery, J.Appl.Phys. 36, 3 (1965).

24. R.C.C. Leite, R.S. Moore,and J.R. Whinnery, Appl.Phys.Lett. 5, 141 (1964).

25. C. Hu and J.R. Whinnery, Appl.Opt. 12, 72 (1973).

26. A. Hordvik, Appl.Opt. 16, 2827 (1977).

27. M.E. Long, R.L. Swofford, and A.C. Albrecht, Science 191, 183 (1976).

28. H.L. Fang and R.L. Swofford, J.Appl.Phys. 50, 6609 (1979).

29. N.J. Dovichik and J.M. Harris, Anal.Chem. 53, 106 (1981).

30. N.J. Dovichi and J.M. Harris, Anal.Chem. 51, 728 (1979).

31. N.J. Dovichi and J.M. Harris, Anal. Chem.52, 2338 (1980).

32. H.W. Latz, H.F. Wyles, and R.B. Green, Anal.Chem.45, 2405 (1973).

33. R.C. Spiker and J.S. Shirk, Anal.Chem. 46, 572 (1974).

34. R.J. Thrash, H. von Weyssenhoff, andJ.S. Shirk, J.Chem.Phys. 55, 4659 (1971).

35. R.A. Keller, E.F. Zalewski, and N.C. Peterson, J.Opt.Soc.Am. 62, 319 (1972).

36. W. Brunner and H. Paul, Opt. Commun. 12, 252 (1974).

37. W. Brunner and H. Paul, Opt. QuantumElectron. 10, 139 (1978).

38. K. Tohma, J.Appl.Phys. 47, 1422 (1976).

39. J.S. Shirk, T.D. Harris, and J.W. Mitchell, Anal.Chem. 52, 1701 (1980).

40. T.D. Harris and J.W. Mitchell, Anal. Chem. 52, 1706 (1980).

LASER EXCITED FLUORESCENCE METHODS IN ANALYTICAL CHEMISTRY

John C. Wright

Department of Chemistry
University of Wisconsin
Madison, Wisconsin 53706

INTRODUCTION

Fluorescence spectroscopy is inherently one of the most sensitive analytical methods. Raman spectroscopy for example has been developed to the point where it is capable of measuring 10^{-3} $\underline{M}$ concentrations. Raman scattering is characterized by cross-sections of ca. 10^{-29} cm^2. Typical values for molecular absorption are 10^{-17} cm^2 so that fluorescence methods should be capable of detecting 10^{-15} $\underline{M}$ concentrations. Such values have in fact been obtained (1) although one is usually limited by background fluorescence before such concentrations are reached (2). Typical values for atomic absorption are 10^{-13} cm^2 so that atomic fluorescence methods should be capable of detecting 10^{-19} $\underline{M}$ or 60 atoms/cm^3 concentrations. A classic experiment by Fairbanks, Hänsch, and Schawlow demonstrated detection of 100 Na atoms/cm^3 in a heated cell (3).

LASER EXCITED ATOMIC FLUORESCENCE

Laser excited atomic fluorescence requires a method for atomizing the sample of interest. Flames, carbon furnaces, and inductively coupled plasmas (ICP) have all been used (4). A sample solution with a concentration of 1 μg/mL (1 part per million) will typically provide 10^{10} atoms/cm^3 in a flame (5). The radiative quantum efficiency in a flame is ca. 0.1 % so that a laser technique capable of detecting 100 atoms/cm^3 in a heated furnace would detect concentrations of 10^{-11} g/mL in the original sample solution. These values are typical of those observed in laser excited atomic fluorescence (6). Graphite furnaces are able to achieve lower detection limits because the atoms in the sample are more restrained to remain within the furnace. Detection limits of 0.5 pg/mL or 2×10^{-14}

gm have been demonstrated (7). The method is highly selective for
specific atoms because of the narrow linewidths in both absorption
and emission. At the present time, atomic fluorescence hasn't ful-
filled a need that cannot be satisfied by more traditional methods
such as ICP spectroscopy but research in the area continues.

MOLECULAR FLUORESCENCE

    Lasers have been extensively studied as a replacement for the
conventional blackbody source in molecular fluorescence. Blackbody
sources work quite well because a reasonable fraction of the emitted
radiation is absorbed by the broad absorption bands of sample
solutions. There are advantages that lasers have nonetheless (8).
Raman scattering from the solvent contributes an appreciable background (1). Since laser excitation is narrow, the Raman scattering
is localized at specific wavelengths so that fluorescence can be
measured between Raman lines. Detection limits of 5×10^{-14} $\underline{M}$ have
been obtained for fluorescence (1). Blackbody sources give detection
limits a factor of 4 worse (9).

    The ability to focus a laser to a small spot size is especially
useful for fluorescence detection of the eluent from liquid chromatography (LC) (8, 10). Two approaches have been used to excite
the eluent without generating interfering fluorescence from
associated windows. A flowing droplet can be suspended at the end
of the LC column with a quartz rod (10). The flowing droplet presents a windowless volume into which an excitation laser can be
focussed. A second approach uses an optical fiber to collect the
fluorescence generated within a glass capillary that is attached
to the end of the column (8). The excitation laser is focused
through the capillary and immediately in front of the fiber optic.
The geometry is such that fluorescence from the capillary walls is
outside the fiber optic's field of view. This approach eliminates
the problem of bubbles forming from the degassing of solvents as
they exit the LC. The combination of fluorescence detection with LC
separation provides a technique that has excellent detection limits
and good specificity for a particular compound of interest. Detection limits are typically in the 10 pg range. For example, concentrations of 1.8 pg/mL of aflatoxin have been successfully detected (10).

    The need for specificity in an analysis is important when
working at low concentration levels where there are many constituents in a real sample. Another method for improving specificity
is to use the selectivity of the antigen-antibody reaction (11).
In order to test for the presence of antibodies in a sample, a mixture of an antigen and a labeled antigen are added. The antigen
is labeled with a fluorescent molecule. Both labeled and unlabeled
antigens bind with the antibodies present. The decrease in the
fluorescence intensity of the labeled antigen reflects the amount

of antibody present in the sample solution. An important step in the measurement is the separation of fluorescence from the labeled antigen and the labeled antigen-antibody complex. Heterogeneous methods for this fluorescence immunoassay rely on achieving a physical separation. Homogeneous procedures rely on spectroscopic differences between the two.

SELECTIVE LASER EXCITATION METHODS

The molecular fluorescence methods discussed thus far obtain their specificity by combination with other techniques. There are a number of other procedures where specificity is achieved spectroscopically. These procedures result in sharp-lined spectra characteristic of the molecule that are compatible with the narrow linewidths of laser sources. The first of these methods used the sharp lines from lanthanide and actinide ions (12-18). The absorption and fluorescence of these ions results from transitions within unfilled $4f^n$ or $5f^n$ orbitals. These orbitals are shielded from outside influences by outer orbitals so their transitions are inherently sharp. Trace analysis of actinide or lanthanide ions is performed by adding $Ca(NO_3)_2$ to a sample solution and coprecipitating them in a CaF_2 precipitate by the addition of NH_4F. The precipitate is recovered and heated to drive off water and introduce small amounts of oxygen. The precipitates are then cooled to cryogenic temperatures and excited with a narrowband tunable laser. Since the absorption lines of the different lanthanide and actinide ions are sharp and distinct, specific ions can be excited so the technique has a great deal of specificity. There is additional specificity in the fluorescence which is also sharp-lined and characteristic of the ion. Typically, concentrations of 10 pg/mL to 20 fg/mL can be detected with these methods (19).

These techniques have also been extended to indirect determinations of nonfluorescent ions (16). High concentrations of lanthanide ions cause clusters of these ions to form in the precipitate. If Er^{3+} is introduced in high concentration, the resulting $Er^{3+}-Er^{3+}$ dimers and higher order clusters have their green fluorescence emission quenched by energy transfer. If other nonfluorescent ions are simultaneously present in much lower concentrations, they can also enter the clusters and form mixed dimers. Thus small concentrations of La^{3+} will result in $Er^{3+}-La^{3+}$ dimers. Energy transfer can no longer occur so the Er^{3+} ions in the mixed dimers will fluoresce. The transitions will be shifted in energy because the different ionic size of the nonfluorescent ions causes different crystal fields so the exact position of the transition is characteristic of the nonfluorescent ion. These indirect methods have detection limits that are comparable with those of the fluorescent ion (Er^{3+} in this example) but they are also very specific to the particular ion.

Organic molecules can also be analyzed with great specifity provided that their spectra can be made narrow. The broad lines that are typical of organic molecules result because molecules can have different conformations and environments in solutions (20). Any individual molecule can have sharp transitions. Four methods have been developed that can select molecules in specific conformations and environments so that sharp-lined spectra are obtained.

The first method involves diluting a sample in a solution that is cooled to liquid helium temperatures to form a glass (21-23). The conventional absorption and fluorescence spectrum of these glasses are broad because the molecules experience a number of different environments and conformations. However if a laser is tuned within the low energy side of the inhomogeneous line profile, only the molecules resonant with the laser will be excited. If the molecules have strong no-phonon transitions, this selection process will cause the resulting fluorescence spectra to have very sharp lines because only a subset of the entire ensemble has been selected. If the laser excitation wavelength is changed, the fluorescence lines will shift in wavelength. This method has been successfully applied to the analysis of polyaromatic hydrocarbons (PAH) (23). The samples are diluted by 10^4-10^5 in glycerol, water, DMSO glasses. Detection limits of 20 pg/mL are achieved in the final glasses for typical PAH compounds like perylene. This limit corresponds to a 0.2 - 2 µg/mL in the sample solution.

A related method uses the Shpol'skii effect. Shpol'skii discovered that when some molecules, particularly PAH's, are frozen in suitable alkane solvents (such as hexane, heptane, etc.), the spectra became very sharp (24-26). The crystal structure of the frozen hydrocarbon forces all the molecules into unique confirmations and environments. Yang, D'Silva, and Fassel have recently succeeded in using this effect for analytical purposes (27, 28). The samples are diluted by 5×10^3 in the alkane solvent and cooled to 4.2 K. A narrowband, tunable laser is then used to selectively excite a particular site of a particular molecule and a high resolution monochromator can be used to measure the sharp-lined fluorescence. Yang and coworkers showed that the technique has a very high specificity. Even closely related molecules could be selectively examined. They were able to obtain detection limits of between 0.1 - 200 ng/mL in the final frozen solution or 0.5 - 1000 µg/mL in the original sample solution.

Matrix isolation methods can provide a more general method of incorporating a sample in a suitable matrix (29-35). The sample of interest is deposited by evaporation onto the cold finger of a liquid helium dewar or a cryogenic refrigerator along with the matrix. The matrix material can be quite different materials such as argon, nitrogen, alkanes, or perfluoro-alkanes. The latter matrices are

particularly favorable for minimizing interactions between the
matrix and sample molecules which are particularly strong if the
sample molecules are polar (33). The line-widths of spectral
transitions depend strongly on the matrix. Alkane matrices form
essentially Shpol'skii systems while argon matrices are more similar
to glasses. Laser induced fluorescence line narrowing can be used
to provide sharp-lined transitions in the disordered matrices. The
selectivity for specific molecules is as high as the Shpol'skii and
line narrowing in glasses methods. Detection limits of 0.4 - 30 pg
are typically obtained.

The last method uses a supersonic jet to achieve narrow lines
(36). A short column that serves essentially as a GC is terminated
with a small orifice through which gas expands into a vacuum chamber.
Samples are injected with a syringe onto the column and helium gas
carries the sample through the column to the jet. The rapid expansion into the vacuum cools the molecules to low temperatures and de-
populates the rotational levels. The molecular spectra become sharp
and well-structured. A laser is crossed with the jet to excite the
molecular fluorescence. The application of supersonic jets to
chemical analysis is new and has only been demonstrated in substances like naphthalene where the transition probabilities are low
(36). Detection limits of 10 ng were obtained for these materials.
There is much room for improvements and detection limits of 1 pg
are predicted.

All of the molecular methods use internal standardization to
quantitate the intensities. A known concentration of another compound that behaves similarly to the compound of interest is introduced into the sample as an internal reference. Sample intensity
measurements are made relative to the internal standard. A particularly attractive internal standardization method was used by Yang
and coworkers (27, 28). They utilized deuterated samples of the
compounds of interest as internal standards. Their selectivity was
sufficiently high that they could resolve the deuterated and un-
deuterated forms. The close match of the properties between the
sample and internal standard improves the accuracy in the quantitation.

CONCLUSIONS

Laser excited fluorescence is an excellent method for a number
of determinations. It can have excellent specificity and very low
detection limits. Fluorescence is restricted however by the requirement that the sample fluoresce. Only a fraction of the organic
molecules fluoresce and only a fraction of those give sharp-lined
fluorescence spectra. Nevertheless, fluorescence methods will become increasingly important because they solve some crucial analysis
problems. A more complete review of this area is available elsewhere (37).

REFERENCES

1. N. Ishibashi, T. Ogawa, T. Imasaka, and M. Kunitake,
      Anal. Chem. 51, 2096 (1979).
2. T.G. Matthews and F.E. Lytle, Anal.Chem. 51, 583 (1979).
3. W.M. Fairbanks,Jr., T.W. Hänsch, and A.L. Schawlow,
      J.Opt.Soc.Am. 65, 199 (1975).
4. J.C. Van Loon, Anal.Chem. 53, 332A (1981).
5. B. Smith, J.D. Winefordner, and N. Omenetto, J.Appl.Phys. 48,
      2676 (1977).
6. J.J. Horvath, J.D. Bradshaw, J.N. Bower, M.S. Epstein, and
      J.D. Winefordner, Anal. Chem. 53, 6 (1981).
7. J.P. Hohimer and P.J. Hargis,Jr., Anal.Chem.Acta 97, 43 (1978).
8. E.S. Yeung and M.J. Sepaniak, Anal.Chem. 52, 1465A (1980).
9. R.J. Kelly, W.B. Dandliker, and D.E. Williamson,
      Anal. Chem. 48, 846 (1976).
10. G.J. Diebold and R.N. Zare, Science 196, 1439 (1977).
11. C.M. O'Donnell and S.C. Suffin, Anal. Chem. 51, 33A (1979).
12. F.J. Gustafson and J.C. Wright, Anal. Chem. 49, 1680 (1977).
13. J.C. Wright, Anal. Chem. 49, 1690 (1977).
14. J.C. Wright and F.J. Gustafson, Anal. Chem. 50, A1147 (1978).
15. F.J. Gustafson and J.C. Wright, Anal. Chem. 51, 1762 (1979).
16. M.V. Johnston and J.C. Wright, Anal. Chem. 51, 1774 (1979).
17. M.V. Johnston and J.C. Wright, Anal. Chem. 53, 1050 (1981).
18. M.V. Johnston and J.C. Wright, Anal. Chem. 53, 1054 (1981).
19. D.L. Perry, S.M. Klainer, H.R. Bowman, F.P. Milanovitch,
      T. Hirschfeld, and S. Miller, Anal.Chem. 53, 1048 (1981).
20. W.C. McColgin, A.P. Marchetti, and J.H. Everly, J.Am.Chem.Soc.
      100, 5622 (1978).
21. I.I. Abram, R.A. Auerbach, R.R. Birge, B.E. Kohler and
      J.M. Stevenson, J.Chem.Phys. 63, 2473 (1975).
22. J.C. Brown, M.C. Edelson, and G.J. Small, Anal.Chem. 50, 1394
      (1978).
23. J.C. Brown, J.A. Duncanson, and G.J. Small, Anal.Chem. 52,
      1711 (1980).
24. E.V. Shpol'skii, A.A. Il'ina, and L.A. Klimova, Dokl.Akad.
      Nauk SSR 87, 935 (1952).
25. E.V. Shpol'skii, Sov.Phys.Usp. 3, 372 (1960).
26. E.V. Shpol'skii, Sov.Phys.Usp. 6, 411 (1963).
27. Y. Yang, A.P. D'Silva, V.A. Fassel, and M. Iles,
      Anal.Chem. 52, 1350 (1980).
28. Y. Yang, A.P. D'Silva and V.A. Fassel, Anal.Chem.53,894 (1981).
29. P. Tokousbalides, E.L. Wehry, and G. Mamantov, J.Phys.Chem.
      81, 1769 (1977).
30. R.C. Stroupe, P. Tokousbalides, R.B. Dickinson, E.L. Wehry,
      and G. Mamantov, Anal.Chem. 49, 701 (1977).
31. P. Tokousbalides, E.R. Hinton, R.B. Dickinson, P.V. Bilotta,
      E.L. Wehry, and G. Mamantov, Anal.Chem. 50, 1189 (1978).
32. J.R. Maple, E.L. Wehry, and G.Mamantov, Anal.Chem. 52,920
      (1980).
33. J.R. Maple and E.L. Wehry, Anal. Chem. 53, 266 (1981).

34. E.L. Wehry and G. Mamantov, Anal. Chem. 52, 643 A (1979).
35. R.B. Dickinson and E.L. Wehry, Anal. Chem. 52, 778 (1979).
36. J.M. Hayes and G.J. Small, Anal. Chem. 54, 1204 (1982).
37. J.C. Wright, Applications of Lasers to Chemical Problems,
       edited by T.R. Evans, John Wiley Interscience (New York,
       1982).

MULTIPHOTON IONIZATION IN ANALYTICAL CHEMISTRY

John C. Wright

Department of Chemistry
University of Wisconsin
Madison, Wisconsin 53706

ATOMIC METHODS USING IONIZATION DETECTION

Multiphoton ionization methods for analysis became attractive when single atom detection was achieved (1-3). Single atoms of Cs formed from the nuclear fission of Cf were detected within a proportional counter after they were ionized by a dye laser tuned to a resonant transition. The ionization continuum was reached after a second photon was absorbed by the excited Cs atom. Since then, the resonance enhanced ionization methods have been extended to include many of the other elements in the periodic table (4). Different schemes are used to cause ionization depending upon the energy level structure of the particular atom. Some atoms require a resonant two photon excitation as the first step while others require two resonant absorptions before the ionization step.

The single atom detection methods work well for samples where individual atoms are easily created in an interference free environment. Most samples of analytical interest however are not easily atomized in a manner compatible with single atom detection. A related ionization method is well-adapted to such measurements (5-16). A conventional burner is used to atomize a sample solution and an electrode is immersed directly in the flame. A potential is applied and the resulting current between the electrode and burner head is measured. If a tunable laser is now directed along the electrode surface and tuned to the resonance frequency of atoms in the sample, the atoms will be excited to energy levels nearer the ionization continuum. Thermal energy in the flame is sufficient to ionize the atoms so an increase in the electrode current will be obtained. The magnitude of the laser enhanced ionization (LEI) de-

pends on how close the resonant level is to the ionization continuum. If it is quite distant, a second laser can be used to excite the atom again to a still higher level (16). The second excitation can improve the detection by several orders of magnitude.

Laser enhanced ionization depends strongly on the electrical
properties of the flame. The presence of easily ionizable atoms
like Na increases the conductivity so that the background currents
are higher (12). The LEI signals must then be observed on top of
the background. In addition, the electric field gradients within
the flame are affected so that the LEI signals from the atoms of
interest are dependent on whether easily ionizable atoms are present. These effects can be held to a minimum if the laser beam is
physically very close to the electrode. This consideration is the
reason why the electrode was placed in the flame (15). The interfering effects are almost eliminated with this geometry. The detection limits and dynamic range of concentrations obtainable by
LEI are as good as those obtained by any of the other flame spectroscopies including laser excited atomic fluorescence. The technique
is also simple to implement and represents a complementary method
to laser excited atomic fluorescence.

RESONANCE ENHANCED MULTIPHOTON IONIZATION IN MOLECULES

The same ideas can be extended to molecular measurement. If a
laser is focused within a gas sample, multiple photon absorption
can cause ionization (17-19). If there is a real molecular level
that is resonant with some combination of the photons, the multiphoton process is enhanced (20-22). In general, there can be
several mechanisms involving different numbers of photons that are
each capable of reaching the ionization continuum since the
ionization continuum is broad and since a molecule has many potential levels available to enhance the ionization. It is difficult
to predict how many photons will be involved in causing ionization
of an arbitrary molecule.

There are a variety of experimental configurations that have
been used to study multiphoton ionization. Both gas bulbs (23) and
supersonic jets (24-28) are used as samples. The supersonic jets
have the advantage that the spectra are sharper and simpler because
of the cooling in the jet. A GC column has also been used as a pre-
separation step for preparing the sample gas (29). After the laser
causes ionization and fragmentation, the ions are detected by proportional counters or electron multipliers (23) or the ions are
mass analyzed with a quadrupole or time-of-flight mass spectrometer (21). The combination of a supersonic jet with a mass
spectrometer provides this method with excellent selectivity for
specific molecules. The excitation transition is usually very
sharp and the mass spectrum of a molecule is characteristic of
the molecule.

The fragmentation pattern is also strongly dependent on the wavelength and the intensity of the laser (20-22, 28). This fact can be viewed either as a complication that will make interpretation that much more difficult or as another source of information about the molecule that will make the selectivity higher. Bernstein and coworkers have found that the wavelength and intensity variations of the fragmentation pattern are not unrelated (30, 31). The important variable is the excess energy imparted to the molecule over that required for the fragmentation and ionization. They demonstrated this dependence by measuring the fragmentation pattern at two different wavelengths and adjusting the laser power so the two fragmentation patterns become equivalent. A more complete theoretical discussion of these points is given elsewhere in this book (32).

Two lasers can be simultaneously used to change the intermediate levels that participate in the ionization process (33). Two independent wavelengths impart even greater specificity but also greater complexity to the measurement. One can also choose different polarizations of the exciting lasers to give information about the intermediate states in the ionization process in the same way that is done for obtaining symmetry information in two photon spectroscopy (34-39).

There have recently been a number of studies that are directed toward the analytical applications of multiphoton ionization spectroscopy but there needs to be much additional work before the capabilities of the method are fully realized (29, 40). Detection limits of 10^7 molecules/cm^3 have been obtained for naphthalene but these were limited by a number of instrumental problems (41). The most important problem is that although the inherent ionization efficiency of multiphoton ionization is high, the pulsed lasers actually used have a duty cycle of $10^{-6} - 10^{-7}$ so most of the molecules escape detection. Pulsed jets could aid in reducing that problem. Klimcak and Wessel have studied GC as a potential partner to multiphoton ionization and have found encouraging results (29).

Siomos and coworkers have demonstrated that the same methods can be used in liquids (42). They detect the ionization produced by a focused laser with a pair of electrodes biased to an electric field of 15 kV/cm directly immersed in the solution. This approach was extended to LC detection by Winefordner and coworkers who compared multiphoton ionization with photoacoustic and fluorescence detection limits (43).

CONCLUSIONS

Multiphoton ionization spectroscopy is a new method which is rapidly expanding. A great deal of research is going on to better understand the physical mechanisms involved and to develop its capabilities for practical applications. The high specificity and

low detection limits characteristic of multiphoton ionization make
this a promising method that can be expected to play an important
role in the future of chemical measurement.

REFERENCES

1.   G.S. Hurst, M.H. Nayfeh, and J.P. Young, Appl.Phys.Lett. 30,
     229 (1977).
2.   G.S. Hurst, M.H. Nayfeh, and J.P. Young, Phys.Rev. 15A, 2283
     (1977).
3.   S.D. Kramer, C.E. Bemis, J.P. Young, and G.S. Hurst,
     Opt. Lett. 3, 16 (1978).
4.   G.S. Hurst, M.G. Payne, S.D. Kramer, and J.P. Young,
     Rev.Mod.Phys. 51, 767 (1979).
5.   R.B. Green, R.A. Keller, G.G. Luther, P.K. Schenck, and
     J.C. Travis, Appl. Phys. Lett. 29, 727 (1976).
6.   J.C. Travis and J.R. DeVoe, Lasers in Chemical Analysis,
     G.M. Heiftje, J.C. Travis, and F.E. Lytle, Eds.,
     Humana Press (New Jersey, 1981).
7.   K.C. Smith, R.A. Keller, and F.F. Crim, Chem.Phys.Lett. 55,
     473 (1978).
8.   W.B. Bridges, J.Opt.Soc.Am., 68, 352 (1978).
9.   E.F. Zalewski, R.A. Keller, and R. Engleman, J.Chem.Phys. 70,
     1015 (1979).
10. J.C. Travis, P.K. Schenck, G.C. Turk, and W.G. Mallard,
     Anal. Chem. 51, 1516 (1979).
11. E. Erez, S. Laui, and E. Miron, IEEE J. Quantum Electron.
     QE-15, 1328 (1979)
12. G.C. Turk, J.C. Travis, J.R. DeVoe, and T.C. O'Haver,
     Anal. Chem. 51, 1890 (1979).
13. R.B. Green, R.A. Keller, P.K. Schenck, and J.C. Travis,
     J.Am.Chem.Soc. 98, 8517 (1976).
14. G.C. Turk, J.C. Travis, J.R. DeVoe, and T.C. O'Haver,
     Anal. Chem. 50, 817 (1978)
15. G.C. Turk, Anal. Chem. 53, 1187 (1981).
16. G.C. Turk, W.G. Mallard, P.K. Schenck, and K.C. Smyth,
     Anal. Chem. 51, 2408 (1979).
17. P.M. Johnson, Acc. Chem. Res. 13, 20 (1980).
18. P.M. Johnson, J. Chem. Phys. 64, 4143 (1976).
19. I.N. Kryazev, Yu.A. Kudryautsev, N.P. Kuz'mina, and
     V.S. Letokhov, Sov. Phys. JETP 49, 650 (1979).
20. D.M. Lubman, R. Naaman, and R.N. Zare, J.Chem.Phys. 72, 3034
     (1980).
21. L. Zandee and R.B. Bernstein, J.Chem.Phys. 71, 1359 (1979).
22. L. Zandee and R.B. Bernstein, J.Chem.Phys. 70, 2574 (1979).
23. D.A. Lichtin, L. Zandee, and R.B. Bernstein, Lasers in
     Chemical Analysis, G.M. Heiftje, J.C. Travis, and
     F.E. Lytle, Eds., Humana Press (New Jersey, 1981).
24. J.H. Brophy and C.T. Rettner, Chem.Phys.Lett. 67, 351 (1979).
25. S. Leutwyler and U. Even, Chem.Phys.Lett. 81, 578 (1981).

26. A. Herrmann, S. Leutwyler, E. Schumacher, and L. Wöste, Chem.Phys.Lett. 52, 418 (1977).
27. D.L. Feldman, R.K. Lengel, and R.N. Zare, Chem.Phys.Lett. 52, 413 (1977).
28. L. Zandee, R.B. Bernstein, and D.A. Lichtin, J.Chem.Phys. 69, 3427 (1978).
29. C.M. Klimcak and J.E. Wessel, Anal. Chem. 52, 1233 (1980).
30. D.A. Lichtin, R.B. Bernstein, and K.R. Newton, J.Chem.Phys. 75, 5728 (1981).
31. J. Silberstein and R.D. Levine, Chem.Phys.Lett. 74, 6 (1980).
32. See chapter by F. Rebentrost.
33. G. Horlick and E.G. Codding, Anal.Chem. 46, 133 (1974).
34. J.O. Berg, D.H. Parker, and M.A. El-Sayed, Chem.Phys.Lett. 56, 411 (1978).
35. D.H. Parker, J.O. Berg, and M.A. El-Sayed, Chem.Phys.Lett. 56, 197 (1978).
36. D.H. Parker and P. Avouris, J.Chem.Phys. 71, 1241 (1979).
37. D.H. Parker, R. Paudolfi, P.R. Stannard, and M.A. El-Sayed, Chem.Phys. 45, 27 (1980).
38. K. Krogh-Jespersen, R.P. Rava, and L. Goodman, Chem.Phys. 44, 295 (1979).
39. M.B. Robin and N.A. Kuebler, J.Chem.Phys. 69, 806 (1978).
40. D.M. Lubman and M.N. Kronick, Anal. Chem. 54, 660 (1982).
41. R. Frueholz, J. Wessel, and E. Wheatley, Anal.Chem. 52, 281 (1980).
42. K. Siomos, G. Kourouklis, and L.G. Christophorov, Chem.Phys.Lett. 80, 504 (1981).
43. E. Voightman, A. Jurgensen, and J.D. Winefordner, Anal. Chem. 53, 1921 (1981).

NONLINEAR SPECTROSCOPIC TECHNIQUES AND THEIR APPLICATIONS TO

ANALYTICAL CHEMISTRY

John C. Wright

Department of Chemistry
University of Wisconsin
Madison, Wisconsin 53706

INTRODUCTION

Coherent nonlinear methods can be understood by examining the relationship between the polarization induced in a molecule and the applied electric field associated with a propagating electromagnetic wave (1-9). At low electric fields, there is a linear relationship with the susceptibility, $\chi^{(1)}$, defining the proportionality constant. High fields can be created by focusing lasers to small areas where the fields become comparable to the binding forces within the atoms or molecules. At some point dielectric breakdown occurs. The induced polarization can no longer follow the electric field linearly and a Taylors series expansion is usually used to approximate the functionality. If one examines the shape of the oscillating polarization at high field intensities, one sees distortions from the sinusoidally varying shape at low fields. These distortions can be described by a Fourier series containing all the harmonics of the fundamental frequency. If one has three lasers at different frequencies, all possible combination frequencies of the three lasers must be considered in describing the distorted polarization. An oscillatory polarization will launch new electromagnetic waves so that in the nonlinear regime, a series of waves will be generated at new frequencies. Each new frequency is associated with a different nonlinear spectroscopy. These new frequencies will be created in all materials since all materials can be polarized by an electric field. Their creation can be particularly efficient if the material has real transitions at the appropriate frequency combinations because then large polarizations can be induced.

81

If two lasers having frequencies of ω_L and ω_S are focused in-to a sample, one of the new frequencies will be $2\omega_L-\omega_S$. This out-put will be resonantly enhanced when ω_L, $(\omega_L-\omega_S)$, or $(2\omega_L-\omega_S)$ matches a natural resonant frequency in a sample. Molecules have vibrational frequencies, ω_V, which can match $(\omega_L-\omega_S)$ and enhance the nonlinear process by about three orders of magnitude. This effect forms the basis of coherent anti-Stokes Raman spectroscopy or CARS. The frequency difference $(\omega_L-\omega_S)$ is scanned by changing either laser while the output frequency $(2\omega_L-\omega_S)$ is monitored. Vibrational resonances show up as increases in the output intensity. Since two photons of ω_L and one photon at ω_S are required, the out-put intensity depends quadratically on the ω_L laser power and line-arly on the ω_S laser power. The output intensity depends quadratic-ally on the sample concentration (1-9).

The generation of light at new frequencies requires that the exciting radiation be phase matched to that generated (1, 6, 10). The wavelength of the oscillating polarization is determined by the indices of refraction at the laser frequencies generating it. The oscillating polarization in turn launches a new wave which has a wavelength determined by the index of refraction at the new fre-quency. Since any material has dispersion, the wavelengths of the new propagating wave will not match the wavelength of the polari-zation responsible for it so that a phase mismatch will develop. To match the wavelengths, the exciting lasers are crossed at a small angle $(1-2^{\circ})$. This matches the wavelengths over a limited region of the sample and allows coherent light to be generated efficiently. It exits the sample as a coherent beam propagating at a different angle from the other beams.

The vibrational CARS resonances are always observed on a back-ground that comes from the nonresonant contribution - i.e., the fact that any material will exhibit an induced polarization (6). This background limits how low a sample concentration can be measured with CARS. There can also be interference between the non-resonant and resonant contributions that affect the shape of the line profiles (5-9). The line shape in CARS is determined by $|\chi^{(3)}|^2$ where $\chi^{(3)}$ is the third order susceptibility expressing the proportionality between the polarization and the cube of the net electric field in the Taylors series expansion. If $\chi^{(3)}$ has reso-nant and nonresonant parts $(\chi^{(3)} = \chi_R^{(3)} + \chi_{NR}^{(3)})$, then $|\chi^{(3)}|^2$ will have cross-terms. The resonant part of the susceptibility will also have a real and an imaginary part since dissipation is possible when the system is driven at resonance. The real part will have a dispersive lineshape at resonance and in combination with $\chi_{NR}^{(3)}$ in the cross-terms gives the resonance a dispersive shape when these cross-terms become important. Thus if the sample concentration is high enough to make $\chi_R^{(3)} >> \chi_{NR}^{(3)}$, the lineshape is a symmetrical Lorentzian but at low concentrations, it changes to a dispersive shape (6).

The resonant part of the CARS susceptibility can be increased 10^3–10^6 times further above the nonresonant background if additional resonant enhancements can be employed. Excited electronic states of molecules allow resonance with ω_L and/or $2\omega_L$–ω_S frequencies. This situation is called resonance CARS (11–19). The detection limits for resonance CARS are lower by 10^3–10^6 as a result. There are some very interesting changes in lineshape that occur as a result of interference effects between the two resonances. These changes are described in several references (14–20).

Raman spectra of molecules that absorb near the laser wavelength are notoriously difficult to obtain because of competing fluorescence. Resonance CARS is often used in these cases because CARS has high fluorescence rejection capabilities (10). The CARS signal is coherent, directional, and well-defined in its spatial position while the fluorescence is incoherent and isotropic. A large part of the fluorescence can be rejected by using an aperture that passes the coherent CARS signal so that the CARS spectrum can be obtained even for highly fluorescent samples.

CARS is also used to probe environments that would be hostile to other measurement probes (6, 9, 21–32). Since the CARS signal is primarily generated within the focal region of the lasers, the measurement can be performed with spatial resolution. A special geometry for phase matching referred to as BOXCARS allows particularly high spatial resolution (33). Phase matching is done by focusing three beams at large angles to each other so the CARS generation occurs over a small crossing region. Since CARS generation is only efficient in the region of crossing, a small volume is examined. This approach is used to map the concentration profiles of species in flames and plasmas. The rotational profiles of molecules can be used to obtain temperature maps as well (6, 25).

The detection limits for CARS are usually set by the fluctuations in the nonresonant background contribution (6, 9). Typically the detection limit for vibrational CARS is 10^{-3} times that of the background component. In liquids, concentrations that are 10^3 times lower than the solvent can be observed. Thus ca. 0.05 $\underline{M}$ concentrations can be seen in water. The same consideration applies for gas phase samples. One of the big differences between gas phase and condensed phase samples is their density. The much lower density of gases makes the CARS generation much less efficient. To compensate for the lower efficiencies, more powerful lasers are usually used.

The detection limits can be lowered further if one can discriminate against the nonresonant background. A very clever discrimination method takes advantage of differences in the polarization characteristics of the resonant and nonresonant parts of $\chi^{(3)}$ (34). The

excitation lasers are aligned so their polarizations are at an
angle to each other. The CARS signal is generated at a still different polarization. An analyzing polarizer blocks the nonresonant part
of the signal but a small part of the resonant part is transmitted.
The detection limits can be lowered by several orders of magnitude
with this approach.

Multiplex CARS was developed to obtain complete CARS spectra
at once (27, 35). Instead of having two narrowband dye lasers which
must be scanned relative to each other, one laser is operated un-
tuned so its output is broad. Then CARS generation is possible over
a range of energies which can be measured using a monochromator.
An efficient detector for the monochromator output is a vidicon
tube.

The output beam at $(2\omega_L-\omega_S)$ can be resonantly enhanced by
other frequency combinations. For example if there are real states
of a molecule at $2\omega_L$ or $(\omega_L+\omega_S)$, the induced polarization will become much larger and give rise to a two-photon process (3). The two
photon process contributes to the same signal as the CARS process
and is usually part of the non-resonant background. Dephasing can
cause a loss of coherence in the two photon process and lead to a
real population of the excited state at $2\omega_L$ or $(\omega_L+\omega_S)$. This dephasing induced population is two-photon absorption.

One of the other frequencies present in the oscillating polar-
ization associated with $\chi^{(3)}$ is at $(\omega_L-\omega_L+\omega_S)$ or ω_S and $(\omega_L-\omega_S+\omega_S)$
or ω_L. They are components at the same frequencies as the driving
lasers. In the absence of relaxation, the number of photons lost at
ω_L is exactly compensated by a gain at ω_L from the oscillating
polarization (36, 37). If there are relaxation paths, there will be
a net loss of photons at ω_L and a net gain at ω_S (assuming $\omega_L > \omega_S$).
Relaxation becomes important when there is a resonance between a
molecular vibrational level at ω_V and $(\omega_L-\omega_S)$. The lasers will drive
the molecule and create a vibrational excitation. A photon will be
lost at ω_L and a photon and a vibration will be created at ω_S and
ω_V respectively. One can then monitor the decrease in the ω_L laser
intensity (Raman loss spectroscopy) or the increase in the ω_S laser
intensity (Raman gain spectroscopy) (36). One can also observe the
heating produced by creating vibrations at ω_V. The latter approach
is called photoacoustic Raman spectroscopy (PARS) (38) or optoacoustic Raman spectroscopy (OARS) (39).

The advantage of these methods over CARS is they do not have
a nonresonant background to limit detection (36). The nonresonant
contribution does not have any relaxation associated with it so
that Raman loss/gain can only be seen for real resonances. There
is a large rejection of interfering fluorescence as well because
the signal is coherent and small apertures can be used to discriminate between the highly directional, coherent laser beams and
the incoherent, isotropic fluorescence.

The detection limits in Raman loss/gain spectroscopy are set
by how well one can measure small changes in the laser intensities.
As described in the chapter on absorption, measurements can approach
shot-noise limited performance (37, 40, 41). Typically one laser is
modulated while the output of the other is monitored at the modulation frequency. Background signals can obscure the Raman signals
if there is some absorption at the wavelength of an amplitude modulated laser. Then thermal effects can occur which will perturb
the other laser at the modulation frequency, even in the absence
of Raman processes. Wavelength modulation is successful in eliminating most of the thermal effects because the laser power is
constant and the thermal effects reach an equilibrium. Changes of
2×10^{-9} in the laser's power have been observed so that $5 \times 10^{-5}\underline{M}$
benzene solutions could be measured (40-43). The detection limits
are low enough to permit the study of Raman scattering from molecules adsorbed onto surfaces, even without enhancement from metal
surfaces (42, 43). Detection limits can be lowered by using a
second resonant enhancement involving excited electronic states.
Resonance Raman gain/loss spectroscopy immediately encounters
strong thermal effects but the Raman signals can still be observed
(44).

In the same manner that multiplex CARS provides simultaneous
observation of CARS at many frequencies, a broadband laser can be
substituted for ω_L or ω_S in Raman loss/gain experiments. If ω_S
is a strong, narrowband laser while ω_L is a broadband laser, absorption lines will appear in the ω_L continuum at the Raman frequencies. This technique is called inverse Raman scattering and is
essentially the same as Raman loss spectroscopy (45, 46).

If a strong laser at ω_L traverses a medium, spontaneous Raman
scattering at ω_V will generate photons at ω_S such that $(\omega_L - \omega_S) =$
ω_V. A few of the photons at ω_S will have sufficient coherence with
ω_L that the two together can drive the vibrations at ω_V to generate
still more photons at ω_S. There is a threshold power for the ω_L
laser above which a coherent beam at ω_S can be generated from
spontaneous Raman scattering. The generation of the new beam is
called stimulated Raman scattering and is essentially the same as
Raman gain spectroscopy (47). It has recently been used to
generate new frequencies from high power lasers with compressed
H_2 gas as the scattering medium (3). H_2 has the highest vibrational
energy, 4155 cm^{-1}, and can produce strong stimulated Raman scattering. High order effects cause a series of new frequencies separated
by 4155 cm^{-1}.

Coherent Raman spectroscopies are still in their infancy and
one can predict many new applications and techniques will be found.
This brief review has illustrated some of the capabilities but the
interested reader can find more extensive reviews elsewhere (1-9).

REFERENCES

1. J.C. Wright, Applications of Lasers to Chemical Problems
        ed. by T.R. Evans, John Wiley Interscience (New York, 1982).
2. P.D. Maker and R.W. Terhune, Phys.Rev. 127, A801 (1965).
3. Y.R. Shen, Rev.Mod.Phys. 48, 1 (1976).
4. R.W. Hellwarth, Prog. Quantum Electron. 5, 1 (1977).
5. M.D. Levenson, Phys. Today, p. 44 (May 1977).
6. W.M. Tolles, J.W. Nibler, J.R. McDonald, and A.B. Harvey,
        Appl. Spectrosc. 31, 253 (1977).
7. A.B. Harvey, J.R. McDonald, and W.M. Tolles, Progress in
        Analytical Chemistry, Vol. 8, Plenum (New York, 1976) p.211.
8. M.W. Tolles and R.D. Turner, Appl. Spectrosc. 31, 96 (1977).
9. A.B. Harvey, Anal.Chem. 50, 905A (1978).
10. R.F. Begley, A.B.Harvey, and R.L.Byer, Appl.Phys.Lett.25,387
        (1974).
11. I. Chabay, G.K. Klauminzer, and B.S. Hudson, Appl.Phys.Lett.
        28, 27 (1976).
12. B. Hudson, W. Hetherington, S. Cramer, I. Chabay, and
        G.K. Klauminzer, Proc.Natl.Acad.Sci.U.S. 73, 3798 (1976).
13. L.A. Carreira, T.C. Maguire, and T.B. Malloy, H.Chem.Phys.
        66, 2621 (1977).
14. L.A. Carreira, L.P. Goss, and T.B. Malloy, J.Chem.Phys. 69,
        885 (1978).
15. L.A. Carreira and L.P. Goss, Advances in Chemistry, A.H. Zewail,
        Ed., Springer-Verlag (Berlin, 1978).
16. P.K. Dutta and T.G. Spiro, J.Chem.Phys.69, 3119 (1978).
17. P.K. Dutta, R. Dallinger, and T.G. Spiro, J.Chem.Phys. 73, 3580
        (1980).
18. R. Igarashi, Y. Adachi, and S. Maeda, J.Chem.Phys. 72, 4308
        (1980).
19. C.M. Roland and W.A. Steele, J.Chem.Phys.73, 5924 (1980).
20. R.T. Lynch, H. Lotem, and N. Bloembergen, J.Chem.Phys. 66,
        4250 (1977).
21. S. Druet and J.P. Taran, Chemical and Biochemisal Applications
        of Lasers, Academic (New York, 1979), p. 187.
22. P. Regnier and J.P.E. Taran, Appl.Phys.Lett. 23, 240 (1973).
23. P. Regnier and J.P.E. Taran, Laser Raman Gas Diagnostics, Plenum
        (New York, 1974), p. 87.
24. J.J. Barrett and R.F. Begley, Appl.Phys.Lett. 27, 129 (1975).
25. F. Moya, S.A.J. Druet, and J.P.E. Taran, Opt. Commun. 13,
        169 (1975).
26. B.R. Hudson, New Applications of Lasers to Chemistry, by
        G. Hieftje, Ed., ACS Symposium Series (Washington, DC, 1978).
27. A.B. Harvey and J.W. Nibler, Appl.Spectrosc.Rev., 14, 101 (1978).
28. L.A. Rahn, L.J. Zych, and P.L. Mattern, Opt.Commun.30, 249
        (1979).
29. G. Laufer, R.B. Miles, and D. Santavicca, Opt.Commun.31,242
        (1979).

30. P. Huber-Wälchli, D.M. Buthals, and J.W. Nibler, Chem.Phys.Lett.
       67, 233 (1979).
31. J.J. Valentini, D.S. Moore, and D.S. Bomse, Chem.Phys.Lett.
       83, 217 (1981).
32. A.B. Harvey, Anal.Chem. 50, 905A (1978).
33. A.C. Eckbreth, Appl.Phys.Lett. 32, 421 (1978).
34. J.L. Oudar, R.W. Smith, and Y.R. Shen, Appl.Phys.Lett. 34,
       758 (1979).
35. W.B. Roh, P. Schreiber, and J.P.E. Taran, Appl.Phys.Lett.
       29, 174 (1976).
36. A. Owyoung and E.D. Jones, Opt.Lett. 1, 152 (1977).
37. A. Owyoung, IEEE J. Quantum Electron. QE-14, 192 (1978).
38. J.J. Barrett and M.J. Berry, Appl.Phys.Lett. 34, 144 (1979).
39. C.K.N. Patel and A.C. Tam, Appl.Phys.Lett. 34, 760 (1979).
40. B.F. Levine and C.G. Bethea, Appl.Phys.Lett. 36, 245 (1980).
41. B.F. Levine and C.G. Bethea, IEEE J.Quantum Electron. QE-16,
       85 (1980).
42. J.P. Heritage, J.G. Bergman, A. Pinczuk, and J.M. Worlock,
       Chem.Phys.Lett. 67, 229 (1979).
43. J.P. Heritage and D.L. Allara, Chem.Phys.Lett. 74, 507 (1980).
44. J.P. Haushalter and M.D. Morris, Anal.Chem. 53, 21 (1981);
       J.P. Haushalter, C.E. Buffett, and M.D. Morris, Anal.Chem. 52,
       1284 (1980).
45. A. Lau, W. Werncke, M. Pfeiffer, K. Lenz, and H.J. Weigmann,
       Sov.J.Quantum Electron. 6, 402 (1976).
46. W. Werncke, A. Lau , M. Pfeiffer, H.J. Weigmann, G. Hunsalz,
       and K. Lenz, Opt. Commun. 16, 128 (1976).
47. M. Maier, Appl. Phys. 11, 209 (1976).

LASER MEASUREMENTS OF TRACE GASES IN THE ATMOSPHERE AND IN THE LABORATORY

W. Krieger
Max-Planck-Institut für Quantenoptik
D-8046 Garching, Fed. Rep. of Germany

H. Walther
Sektion Physik, Universität München, and
Max-Planck-Institut für Quantenoptik
D-8046 Garching, Fed. Rep. of Germany

SUMMARY

A survey of laser investigations in the atmosphere is given, and more recent results of air pollution measurements by the differential absorption method are discussed in detail. In a second part single-atom detection in the laboratory is reviewed, and as an example an experiment is presented in which collision processes of single atoms were studied.

In the past decade the detection of small concentrations of atoms and molecules in the atmosphere and in the laboratory has been improved tremendously by the application of lasers. Due to the small divergence of the laser beam these methods are applicable over large distances so that for example the continuous monitoring of the ozone layer from ground-based stations became feasable. Using the high spectral density of laser light single-atom detection has been demonstrated in the laboratory. This is an important progress and implies many applications as e.g. in isotopic dating or for the production of extremely pure materials. Essential for both fields is the wavelength tunability of lasers since all optical detection methods rely on the absorption and emission of specific wavelengths by atoms and molecules.

1. LASER INVESTIGATIONS IN THE ATMOSPHERE

For the analysis of gases in the atmosphere with the help
of lasers either the absorption or the scattering of laser beams
can be measured.[1] Absorption measurements yield the highest sensitivity and are very simple, but give concentrations only inte-
grated along the path of the light.[2] To cover a larger area by
the laser beam, mirrors can be used. A very simple setup uses
topographic targets for reflection of the light back to its
source for detection. By the use of mirror arrays, absorption
measurements can yield, like in tomography, the spatial distribution of gas constituents.[3]

Normally, absorption measurements can be performed with low-
power lasers (e.g. diode lasers). Even arc lamps with spectral fil-
tering can be applied. Such a setup was very successful in detecting SO_2, N_2O, CH_2O, O_3, NO_2 and NO_3 with very high sensitivity.[4]

A remarkable improvement of the detectivity in absorption
measurements becomes possible by the heterodyne technique, where
a tunable laser is used as local oscillator and a photodetector
as mixer.[5] The signal at the intermediate frequency is amplified by a suitable amplifier of narrow bandwidth. The heterodyne
method is already being used in satellite- or balloon-borne experiments for detecting atmospheric gases (e.g. O_3) by absorption
with the skylight serving as light source.[6]

Absorption measurements, although of highest sensitivity,
nevertheless have the disadvantage of delivering information
about the concentration distribution only under certain re-
stricting conditions. Studies making use of atmospheric scattering are free from this disadvantage since time-dependent
observation of the backscattered light from a pulsed laser
allows a spatial resolution (as in RADAR). Since light re-
places the radio waves, these methods are known under the
name of LIDAR.

Four different light scattering mechanisms are observed in
the atmosphere: Mie scattering, Rayleigh scattering, Raman scattering and fluorescence scattering. At lower altitudes Mie scattering from aerosols, clouds and dust particles is dominant.
Rayleigh scattering from the molecular constituents of the atmosphere is about two orders of magnitude smaller. It becomes dominant above heights of 30 km, where virtually no aerosols exist.
Raman scattering and fluorescence scattering allow to measure
atmospheric components selectively. Due to the small Raman cross
sections, however, the Raman signals of typical pollutants are
about nine orders of magnitude lower than the Rayleigh signal to
which all gases contribute. From this value it is obvious that
Raman scattering is useless for the remote detection of gases at
lower concentrations.

Scattering cross sections for fluorescence scattering are several orders of magnitude larger. In the lower atmosphere problems arise from collisional quenching of the fluorescence: in general, only 1 % of the atoms can radiate their excitation energy, whereas the majority transfers this energy to the molecules of the air by collisions.

The detection of atoms in the higher atmosphere by fluorescence radiation is, however, much easier: a LIDAR experiment of this kind was performed by Sandford et al.[9] for the first time. Na atoms were detected in these measurements; other alkaline metals were found in later ones.

A (generally double-structured) layer about 15 km thick was found at an altitude of 90 km, the Na atom density being of the order of 10^3 cm^{-3}. The laser experiments clearly established a coincidence between the sodium concentration and the occurrence of meteorites. For the formation of atomic sodium, the ratio of the concentrations of atomic oxygen to ozone is essential, as the following reactions take place:

$$NaO + O \rightarrow Na + O_2$$
$$Na + O_3 \rightarrow NaO + O_2$$
$$Na_2 + O \rightarrow NaO + Na$$

The reaction of atomic sodium and atomic oxygen proceeding via a triple collision can be neglected at the low densities. In the first and third reactions, Na atoms are produced in the excited ^2P state. Probably these reactions contribute to the aurora. Atomic oxygen is produced by photodissociation of O_3 and O_2. Obviously, the O/O_3 ratio is of a value ensuring the survival of atomic sodium just at an altitude of 9o km.

As mentioned above, the observation of absorption excludes measurement of the density distribution of a gas. This disadvantage can be overcome by using Mie scattering as a "mirror". The position of this "mirror" can be determined by the time elapsing between the emission of the laser pulse and the detection of the backscattered signal. The dependence on wavelength of the Mie scattering is low. It is therefore possible to eliminate the local variation of the scattering by measurement at two different wavelengths. Only one of these has to be absorbed by the gas to be detected. Both wavelengths must be sufficiently close together, to ensure that their Mie scattering is in fact equal.

This method, called "differential absorption", was first applied by Rothe et al.[10] for measuring air pollutants. Theoretical studies[11] had shown before that it is the most sensitive of all LIDAR methods.

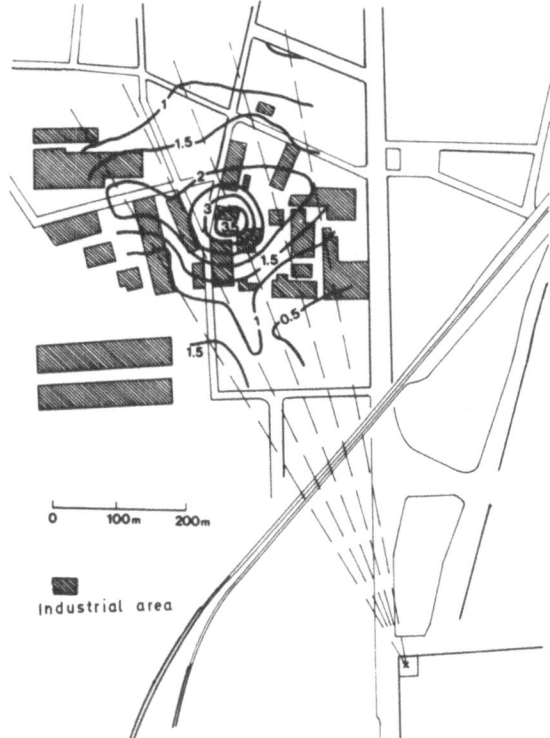

Fig. 1. Distribution of NO_2 concentration above a chemical
plant. Concentrations are given in ppm. The results
were obtained by averaging about 40 000 laser shots.[10]

In the first measurements[10] the apparatus was equipped
with a tunable dye laser. The concentration of NO_2 near a
chemical factory was measured. With only 1 mJ of pulse energy,
concentrations as low as 0.2 ppm could be detected at distances
of 4 km. The distribution of the NO_2 concentration above the
plant was determined by varying the direction of the laser
beam. Connecting points of equal concentration measured in
five different directions, the map of iso-lines shown in Fig. 1
was obtained. It becomes evident from this picture which building is the source of the pollutant.

By means of a dye laser the differential absorption method
can also be applied to SO_2 and O_3. SO_2 measurements were carried
out e.g. by Svanberg et al.[8] Measurements of the ozone concentration in the stratosphere will be discussed here in the following.

After Rowland's and Molina's publication[14] of the catalytic cycle:

$$Cl + O_3 \rightarrow ClO + O_2$$
$$ClO + O \rightarrow Cl + O_2$$

(which, by the catalytic action of Cl atoms, converts O_3 into O_2) the ozone problem has gained much attention. From chlorinated fluoromethanes (for example, used in spray cans) chlorine is produced by photodissociation:

$$CFCl_3 \xrightarrow{h\upsilon} CFCl_2 + Cl$$
$$CF_2Cl_2 \xrightarrow{h\upsilon} CF_2Cl + Cl$$

As virtually no sinks exist for the chlorofluoromethanes, a steadily increasing chlorine concentration which may lead to a destruction of the ozone layer must be taken into account. A permanent observation of the ozone layer is therefore as necessary as it is useful.

First measurements of the ozone layer with dye lasers were performed by Gibson et al.,[9] and later by Megie et al.[12] and Chanin et al.[13] As Mie scattering can be neglected in the upper atmosphere, Rayleigh scattering was used as "mirror" in these experiments.

In these laser measurements time intervals in the range of several hours were required to get an accuracy of 30 %. This also holds for the measurement of Uchino et al.;[15] they used a XeCl laser at 308 nm without having a second frequency necessary as a reference line for the differential-absorption technique. Instead of this they used the data of a balloon sonde, launched the next day, as a reference.

In Munich we have constructed a new setup for ozone measurements in the troposphere taking into account the requirements for a high power laser as well as for a proper reference line. We use a XeCl laser with a pulse energy of 130 mJ and a repetition rate of up to 100 Hz. Its radiation is focussed into a high-pressure methane cell to generate the reference line by stimulated Raman scattering. The backscattered light is focussed by a spherical mirror. The two wavelengths are separated by a dichroic beam splitter and two interference filters. The photons backscatterted from the stratosphere are counted, and the corresponding count-rates stored every 66.7 ns and afterwards transferred to a computer for further data evaluation.

The major advantage of this setup is that simultaneous measurements at both wavelengths are possible, thus eliminating all problems associated with rapidly changing atmospheric conditions (turbulences). Owing to the high repetition rate and pulse energy of the XeCl laser the apparatus is expected to give ozone profiles with an accuracy of a few percent in a measuring period of 10 min.

One disadvantage of all ground-based stratospheric LIDAR measurements is the rather strong decrease of the light intensity in the first few kilometers primarily due to Mie scattering, and in the UV spectral region also due to Rayleigh scattering. The measurements with our ozone LIDAR are therefore being made from the summit of the Zugspitze in the Alp Mountains (altitude 3 km). This difference in altitude gives rise to an increase of the intensity of the stratospheric backscattering by a factor of about three for clean air, and of twenty for hazy air.

When the differential absorption method is to be extended to a larger number of different pollutants, measurements are to be carried out in the infrared range of the spectrum. The most universal setup would be obtained by using a continuously tunable laser. As such lasers, at present, are still difficult to be used in field measurements, molecular lasers (e.g. DF, HF, CO, CO_2 and N_2O lasers) must be employed and measurement must rely on accidental coincidences of laser emission lines with absorption lines of the pollutants.[1,2,16,17] Several pollutants can even be detected simultaneously by a multi-line measurement or sequential measurements using different emission lines.

In the following, some results with a setup equipped with a multi-gas laser are reported. Details of the apparatus are published elsewhere.[2]

The ethylene concentration around an oil refinery near Ingolstadt was measured. The ethylene detected leaks out of the distillation plant. These measurements were made with a CO_2 laser placed at a distance of about 500 m from the refinery. An example is shown in Figs. 2a and 2b.

For environmental surveillance not only is the distribution of the pollutant interesting, but also the amount emitted per unit time. This quantity can be determined by comparing the distribution obtained from a differential absorption measurement with a diffusion model that takes turbulent diffusion into account. For the given example the rate of emission of ethylene was determined to be 0.95 g/s.

The method used here needs to be tested further; however, it can be said that, in principle, it is useful for determining rates of emission.

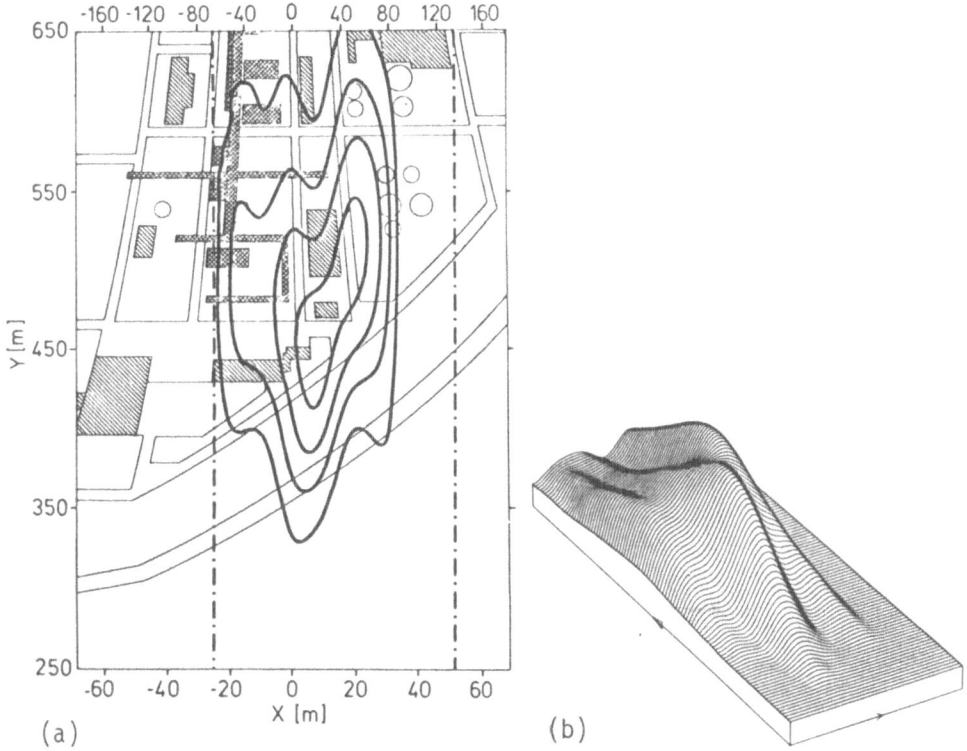

(a)   (b)

Fig. 2.   (a) Map of the refinery and contour lines of ethylene
              concentration. The lines start with 20 ppb, the
              inward increase being 20 ppb.
          (b) Ethylene concentration (vertical axis) above the
              refinery.

2. SINGLE-ATOM DETECTION IN THE LABORATORY

Using lasers the detection of small amounts of substances
in the laboratory has been improved in the past few years to the
point where the detection of single atoms has become possible and
was achieved in several laboratories.[20] Single-particle detection
is widely used in nuclear physics applying radioactive decay. The
detection of single neutral and stable atoms, however, got only
possible after frequency tunable lasers became available.

As in atmospheric studies lasers are used in this field to
selectively excite specific atoms. One way to detect the excited
species is the observation of the subsequently emitted fluorescence

light. This method has e.g. been applied to measure atomic densities down to 100 atoms/cm^{-3} [21] and to detect single trapped ions.[22]
Due to the efficiency limits in collecting and detecting photons,
however, the atoms have to go through the excitation-emission cycle
more than once during their stay in the interaction region with the
laser radiation. Therefore long interaction times and cw lasers are
necessary which restrict the fluorescence-detection method considerably.

A second, more successful detection scheme makes use of the
fact that single ions or electrons can be collected and detected
with an efficiency approaching unity with the help of secondary
electron multipliers or proportional counters. The laser is used
to ionize specific atoms which can be accomplished in a variety
of ways. Usually in a first step the desired atoms are resonantly
excited with a tunable laser. Further laser photons of the same
or different energy then take the excited atoms to the ionization
continuum or to autoionizing states which have larger excitation
cross sections. Ionization schemes have been indicated for nearly
all elements in the periodic table.[20] The important point is
that the output intensities of existing lasers are sufficient to
ionize all atoms of a specific kind in a reasonably large volume.

In a different scheme tunable lasers are used to excite
Rydberg states close to the ionization limit in several steps.
From these states the atoms are quantitatively ionized by collisions in a buffer gas or by applying a weak electric field. As
this ionization scheme usually uses more than one resonant process
its efficiency and selectivity are increased.

First successful experiments to detect single atoms were done
by Hurst et al.[23] who evaporated Cs atoms into a proportional
counter which was equipped with windows to transmit the laser beams.
The atoms diffused into the laser interaction zone where they were
ionized and subsequently counted. A selectivity of one Cs atom in
10^{19} buffer gas atoms was achieved. In further experiments single
Cs atoms from photodissociated CsI and from fission decay of ^{252}Cf
were observed.[24] The high buffer gas pressure in proportional
counters, however, leads to considerable line broadening and thereby decreases the selectivity especially when different isotopes are
involved.

This drawback is overcome by evaporating the atoms in a vacuum
chamber and deflecting the ions or electrons into a secondary electron multiplier by means of which they are counted. The selectivity
of this method can be improved by inserting a quadrupole mass filter.
Single Xe atoms of a selected isotope were counted in this way.[25]

Using the excitation of Rydberg states followed by field
ionization single Na and Yb atoms were detected.[26] From these

experiments it was concluded that Yb isotopes with a relative content of 10^{-10} to 10^{-15} may be detected.[27]

Single-atom detection is, however, not only important for various analytical applications. The possibility of "counting" or "following" single atoms opens up new experiments in fundamental physics. In the following an experiment will be described which allows to observe collisions of single atoms.

The setup[28] consists of a proportional counter arrangement as described by Hurst et al.[23] In contrast to their experiment, however, excitation to Rydberg states and subsequent ionization by collisions with the rare-gas atoms in the counter was used. The two-step resonant excitation to a Rydberg state has now the advantage that the second excitation step can be tuned to match a transition from an intermediate state which is different from the level populated in the first excitation step. This is illustrated for our experiment on Na atoms in Figs. 3 and 4. One laser is tuned to the $3^2P_{1/2}$ transition, and the second to a line starting from $3^2P_{3/2}$ to a Rydberg state. It is clear that a signal in the proportional counter is only observed when energy changing collisions ($3^2P_{1/2} \rightarrow 3^2P_{3/2}$) occur immediately after laser excitation.

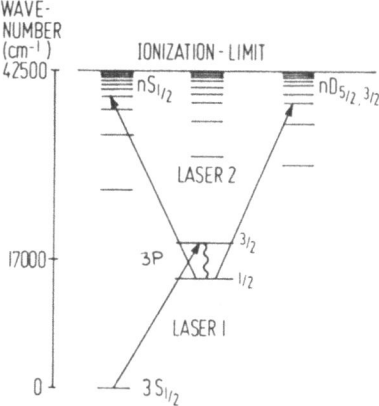

COLLISIONAL ENERGY TRANSFER
IN THE 3P STATES

Fig. 3. Excitation steps used to populate the Rydberg states.
A signal pulse is observed in the proportional counter
when energy-changing collisions occur in the intermediate state indicated by the wavy line.

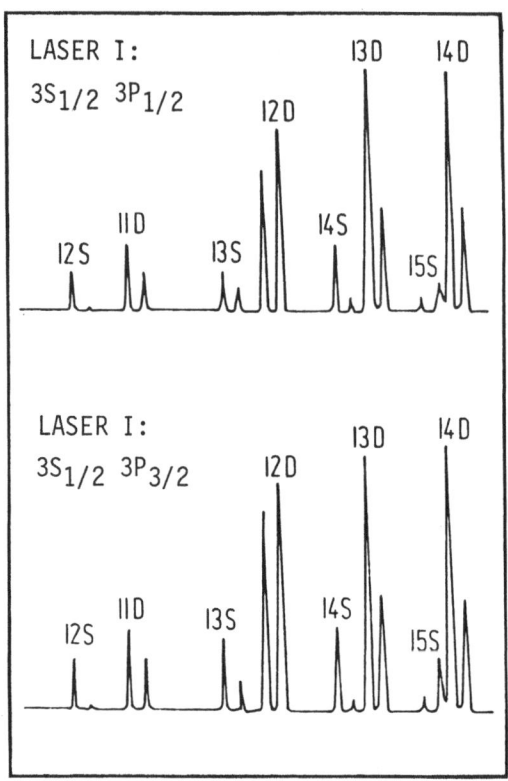

Fig. 4. Signal in the proportional counter when the laser
         performing the second excitation step is tuned. The
         first excitation step is performed to different fine
         structure states in the upper and lower signal curve.

Since the setup provides a one-atom sensitivity as was
demonstrated in various experiments it is possible to follow
the collisional behaviour of a single atom e.g. to measure the
average time between excitation and collision, the dependence
of the collisions on alignment or orientation; it is even possible to get an estimate for the duration of a collision.
These are interesting new possibilities for further new experiments.

REFERENCES

1. The subject has been reviewed in several books, e.g.
    "Laser Monitoring of the Atmosphere", D. Hinkley, ed.,
    Topics in Applied Physics Vol. 14, Springer Verlag
    Berlin, Heidelberg, New York (1976)
    For a compilation of very recent results see:
    D. K. Killinger, A. Mooradian, eds.: "Workshop on Optical
    and Laser Remote Sensing", in print
    Springer Series in Optical Sciences, Springer Verlag
    Berlin, Heidelberg, New York (1982)
2. K. W. Rothe, H. Walther in: "Tunable Lasers and Applications,
    A. Mooradian, T. Jaeger, P. Stokseth, eds., Springer
    Series in Optical Sciences Vol. 3, Springer Verlag
    Berlin, Heidelberg, New York (1976)
3. R. L. Byer, L. A. Shepp, Optics Lett. 4:75 (1979)
4. U. Platt, D. Perner, H. W. Patz, Journ. Geophys. Rev.
    84:6329 (1979); D. Perner, U. Platt, Geophys. Rev.
    Lett. 6:917 (1979); U. Platt, D. Perner, A. M. Winer,
    G. W. Harris, J. N. Pitts. Jr., Geophys. Rev. Lett.
    in press
5. R. T. Menzies, M. S. Shumate, Science 184:570 (1974)
6. R. T. Menzies in: "Laser Spectroscopy III", J. L. Hall and
    J. L. Carlsten, eds., Springer Series in Optical
    Sciences, Springer Verlag Berlin, Heidelberg, New
    York (1977)
7. V. E. Derr, M. J. Post, R. L. Schwieso, R. F. Calfee,
    G. T. Mc Nice, A theoretical Analysis of the Infor-
    mation Content of LIDAR Atmosphere Returns, NOAA
    Technical Report ERL 296 - WPL 29
8. S. Svanberg in: "Surveillance of Environmental Pollution
    and Resources by Electromagnetic Waves - Principles
    and Applications", T. Lund, ed., Nato Advanced Study
    Institute Series, D. Reichel Publishing Company,
    Dordrecht, Holland (1978)
9. M. C. Sandford, A. J. Gibson, J. Atmosph. Terr. Phys.
    32:1423 (1970)
    A. J. Gibson, L. Thomas, Nature 256:561 (1975)
10. K. W. Rothe, U. Brinkmann, H. Walther, Appl. Phys. 3:114
    (1974)
    K. W. Rothe, U. Brinkmann, H. Walther, Appl. Phys.
    4:181 (1974)
11. R. L. Byer, M. Garbuny, Appl. Optics 12:1496 (1973)
12. G. Megie, J. Y. Allain, M. L. Chanin, J. E. Bamont,
    Nature 270:329 (1977)
13. M. L. Chanin, J. W. Allain, J. Pelon, Ninth International
    Laser-Radar Conference, Munich, July 1979
14. M. J. Molina, F. S. Rowland, Nature 249:810 (1974)
15. O. Uchino, M. Maeda, M. Hirono, IEEE, QE-15, 10:1094 (1979)

16. J. C. Petheram, Applied Optics 20:3951 (1981)
17. D. K. Killinger, N. Menyuk, IEEE J. of Quantum Electronics,
    QE-17, 1917 (1981)
18. M. S. Shumate, R. T. Menzies, W. B. Grant, D. S. Dougal,
    Applied Optics 20:545 (1981)
19. W. Baumer, Ph. D. Thesis, Fakultät für Physik, Ludwig-
    Maximilians-Universität München, July 1979
20. For a review, see:
    G. S. Hurst, M. G. Payne, S. D. Kramer, and J. P. Young,
    Rev. Mod. Phys. 51:767 (1979),
    V. I. Balykin, G. I. Bekov, V. S. Letokhov, and V. I. Mishin,
    Sov. Phys. Usp. 23:651 (1980)
21. W. M. Fairbank Jr., T. W. Hänsch, and A. L. Schawlow,
    J. Opt. Soc. Am. 65:199 (1975)
22. W. Neuhauser, M. Hohenstatt, P. Toschek, and H. Dehmelt,
    Phys. Rev. Lett. 41:233 (1978)
23. G. S. Hurst, M. G. Payne, M. H. Nayfeh, J. P. Judish, and
    E. B. Wagner, Phys. Rev. Lett. 35:82 (1975)
    G. S. Hurst, M. H. Nayfeh, and J. P. Young, Phys. Rev. A
    15:2283 (1977)
24. L. W. Grossman, G. S. Hurst, M. G. Payne, and S. L. Allman,
    Chem. Phys. Lett. 50:70 (1977)
    S. D. Cramer, C. E. Bemis Jr., J. P. Young, and G. S. Hurst,
    Opt. Lett. 3:16 (1978)
25. C. H. Chen, G. S. Hurst, and M. G. Payne, Chem.Phys. Lett.
    75:473 (1980)
26. G. I. Bekov, V. S. Letokhov, and V. I. Mishin, JETP Lett.
    27:47 (1978)
    G. I. Bekov, V. S. Letokhov, O. I. Matveev, and V. I. Mishin,
    Opt. Lett. 3:159 (1978), Sov. Phys. JETP 48:1052 (1978)
27. G. I. Bekov, E. P. Vidolova-Angelova, V. S. Letokhov, and
    V. I. Mishin in: "Laser Spectroscopy IV", H. Walther, and
    K. W. Rothe eds., Springer, Berlin, 1979
28. W. Ohnesorge, Diploma-thesis, University of Munich,
    unpublished.

PART III

SPECTROSCOPIC AND DYNAMICAL STUDIES

VUV LASER SPECTROSCOPY OF ATOMIC AND MOLECULAR HYDROGEN

K. H. Welge

Fakultät für Physik
Universität Bielefeld
4800 Bielefeld, Fed. Rep. of Germany

INTRODUCTION

Considerable progress has recently been made in the development of techniques for the production of tunable laser light in the vacuum ultraviolet down to, and below, 1000 Å. Vuv can be generated efficiently with pulsed dye laser by stimulated anti-Stokes Raman scattering in hydrogen to ~ 1500 Å[1] while the whole vuv range to ~ 900 Å can be covered by frequency tripling and mixing in gases and metal vapors[2]. With commonly used Nd:YAG or Excimer pumped pulsed laser systems operated typically at 5 - 10 nsec pulse duration and 10 Hz repetition rate the latter techniques yield intensitites from 10^9 to 10^{13} photons/pulse, depending on details of the pumping and frequency conversion procedure. Table 1 summarizes essential experimental parameters of present tunable vuv laser light generation by frequency tripling and mixing. Synchrotron radiation parameters are also given for comparison. The table is meant to provide rough, order-of-magnitude indication only.

The generation of tunable vuv laser light has evidently reached a state where the full potential of laser spectroscopy can be exploited in this wavelength region. As to chemistry, applications will be in spectroscopic analysis selective and sensitive detection, and state selective excitation of chemical processes, such as dissociation and ionization.

In this paper we briefly review some recent experiments with tunable vuv laser performed in our laboratory with hydrogen atoms and molecules, both species of fundamental and wide interest in

103

Table 1: Tunable vuv generation by synchrotron and laser frequency conversion (tripling and mixing). Order-of-
magnitude indication of some properties at present
state.

SYNCHROTRON - LASER (tunable)

	Synchrotron	Laser
.. spectral range	unlimited	$\geq$ 1000 Å ($\geq$ 750 Å)
.. tuning range	unlimited	1 - 100 Å
.. directional prop.	m rads	m rad
.. time structure:		
pulse duration	10 ps - 1 ns	ps - 10 ns - (cw)
rep. rate	$\geq 10^6$ Hz	single - 100 Hz
.. intensity:		
pulse	$\lesssim 10^6$/pulse/Å	10^{10} - 10^{12} - 10^{14}/p
cw	-	10^5/sec
average	$\sim 10^{12}$/sec/Å(*)	10^{11} - 10^{13} - 10^{15}/sec
.. resolution		
($\lambda/\Delta\lambda$)	$\sim 10^2$ at 10^{12} ph/s	$\sim 5\times10^4$ - 10^7
	$\sim 5\times10^5$ at 10^6 ph/s	$\sim$ 30 - 0.1 GHz
		($\sim$ MHz with cw).
.. polarization	parallel; elliptical	parallel; circular ($\geq$ 1130 Å)

*) without wiggler

chemistry and physics. The main objective of the paper is to provide an indication of the potential and the present state-of-art
of tunable laser application in the far vacuum ultraviolet. As
vuv laser spectroscopy is still in a relatively infant state, the
experiments have partly the character of feasibility studies.

HYDROGEN ATOM

One-photon excitation of the hydrogen atom to the first
state n=2 with tunable vuv laser light around 1216 Å (Lyman-α
line) has been achieved previously[3,4]. By non-resonant frequency
tripling in krypton intensities at Lyman-α of the order of 10^{10}
photon/pulse with 5-10 nsec pulse duration are commonly obtained[2]. Higher intensities up to $\sim 10^{12}$ photons/pulse can be produced by the tripling process with optimisation of the phase matching conditions[5] or by resonant two-photon amplification in the
frequency mixing technique[6].

Two-photon ionization

Two-photon ionization with resonant excitation at Lyman-α as

first step is a possibility for selective and sensitive detection
of hydrogen atoms. Fig. 1 shows the first two-photon excitation-
ionization spectrum of atomic hydrogen,

$$H(1^2S) + \hbar\omega_{vuv} \rightarrow H(2^2P) + \hbar\omega_{vuv} \rightarrow H^+ + e, \qquad (1)$$

previously obtained under bulk condition with hydrogen of natural
isotopic composition at room temperature[4]. As to the detection
sensitivity, the sensitivity limit defined by a signal-to-noise

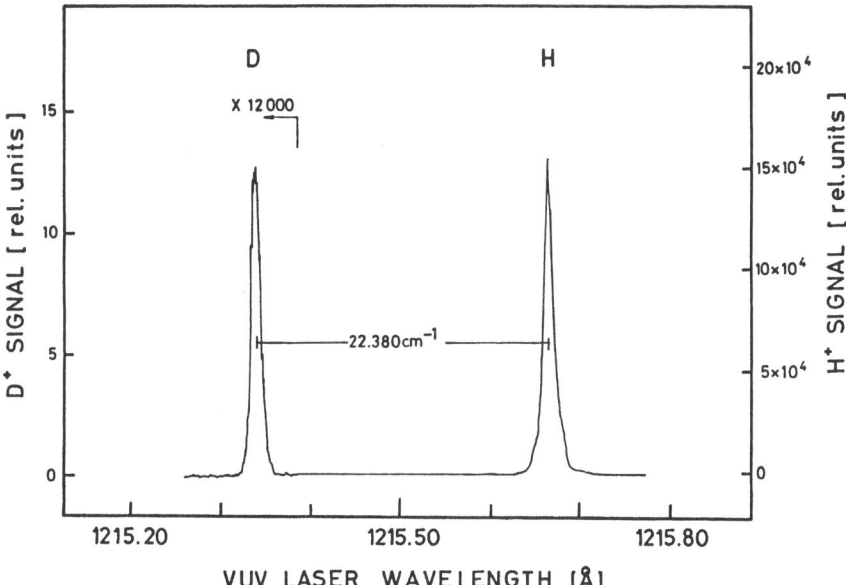

Fig. 1 Two-photon excitation-ionization of atomic hydrogen of
natural isotopic composition from a microwave discharge.

ratio 1:1 and 10 sec signal integration time constant was of the
order of 10^5 atoms/cm^2 in this experiment, where the vuv intensity was ~ 10^9 photons/pulse and the ion collection efficiency
was 10%. The overall spectral resolution was ~ 40 GHz, determined
by the vuv laser bandwidth (~ 30 GHz) and the Doppler width of
~30 GHz at 1216 Å for H atoms at room temperature. With 10^{12} pho-
ton/pulse vuv intensity 100% efficiency and readily attainable
the sensitivity limit will be ~ 10 atoms/cm^3. The laser bandwidth
is given by the method used to produce the uv fundamental for the
Lyman-α at ~ 3648 Å. In this experiment we have added in a crystal the Nd:YAG fundamental (1.06 μ) and the tunable uv radiation
from a dye laser system, an oscillator with 3 amplifier
stages[7,8]. While the dye laser radiation was narrowed sufficiently by known procedures[7,8], the bandwidth of the 1.06 μ line was
the main contribution to the vuv line width. More details may be
taken from ref. 4.

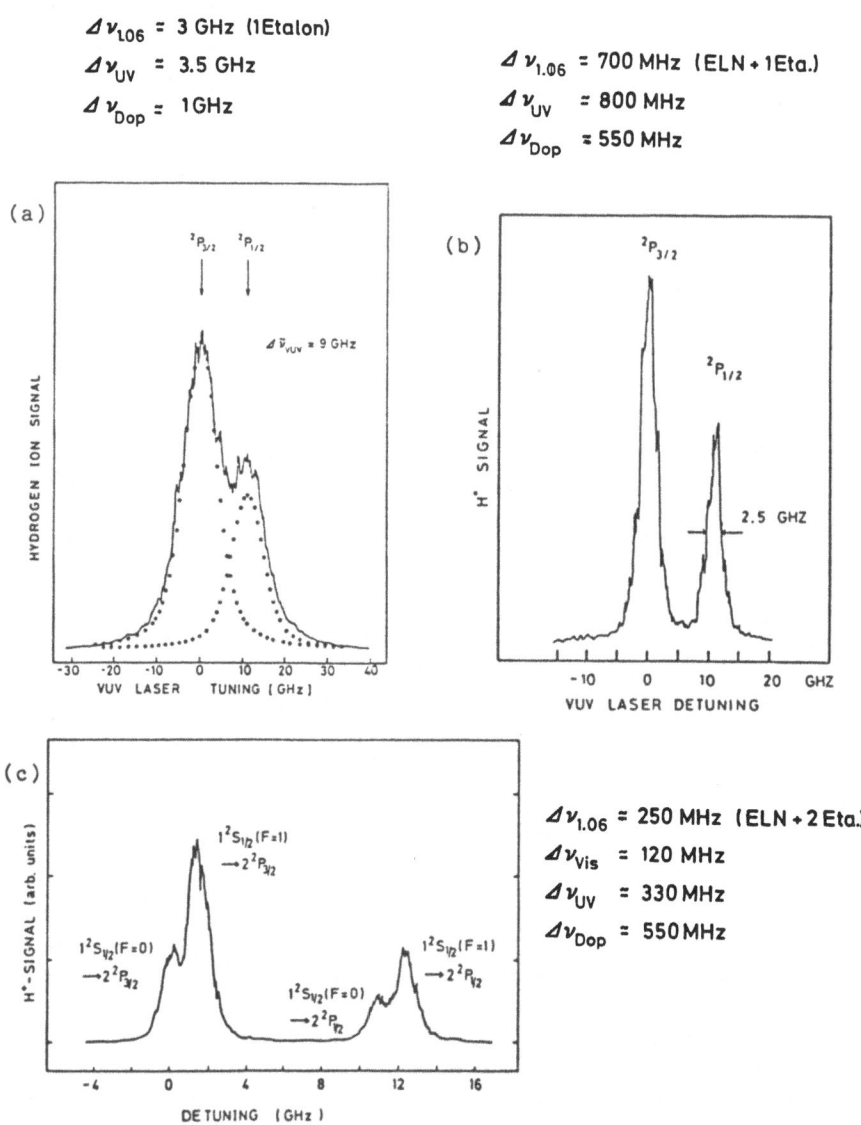

Fig. 2: Two-photon excitation-ionization of the atom in atomic
         beam with different vuv laser bandwidth. Spectrum c
         taken with circularly polarized vuv.

Fig. 2 shows excitation-ionization spectra of the H(1S→2P) transition with higher overall resolution. Different from the previous experiment[4], they have been obtained in a crossed laser-atom beam arrangement to reduce the Doppler width. The overall resolution in the spectra was given mostly by the laser bandwidth which depended mostly on the narrowing procedure used to narrow down the dye laser light and the 1.06 μ line. In the spectrum (c) with highest resolution the hyperfine splitting of the H(1S$_{1/2}$) ground state (1.42 GHz) can be recognized indicating an overall resolution of about 2 GHz. In this measurement the Doppler contribution was ~ o.5 GHz and that of the vuv laser bandwidth ~ 1.5 GHz. This bandwidth was mostly determined by that of the 1.06 μ line which remained when the Nd:YAG oscillator was operated with two etalons and an electronic narrowing device (Quanta Ray). More details may be taken from ref. 9.

Doppler Spectroscopy

The tunability of vuv laser light can be employed for line profile measurements at Lyman-α. As has been shown previously, such Doppler spectroscopy experiments allow to investigate the dynamics of scattering processes[10], including reactions. The applicability of Doppler spectroscopy to hydrogen atoms has been shown and investigated in a previous work where the photodissociation HI → H + I at 266 nm has been used as a first example[11].

Another example presented here is the dissociation of the formaldehyde molecule. Isotopically labeled formaldehyde, HDCO, was used to demonstrate the selectivity of the method. The dissociation was carried out in this feasibility study for convenience at 266 nm, the fourth harmonic of the Nd:YAG laser. At this wavelength the dissociation channels H + DCO and D + HCO and H + D + CO are energetically possible. As in the previous work[11] resonance fluorescence excitation has been employed to measure the lyman-α profiles of H and D.

Fig. 3 shows the profils obtained under collision freee conditions, given by the formaldehyde pressure of 20 mTorr and the delay time of 25 nsec. between the dissociation and the probe laser pulses. The angle θ between the electric field vector of the dissociation light beam and the probe beam direction was θ = 0° in this measurement. Measurements at θ = 45° and 90° yielded the same line profiles.

The measured profils have within experimental precision Gaussian shape. This, together with the independence of the profile on the polarization angel θ straightforwardly indicates that the atoms are produced with Boltzmann-like velocity distribution,

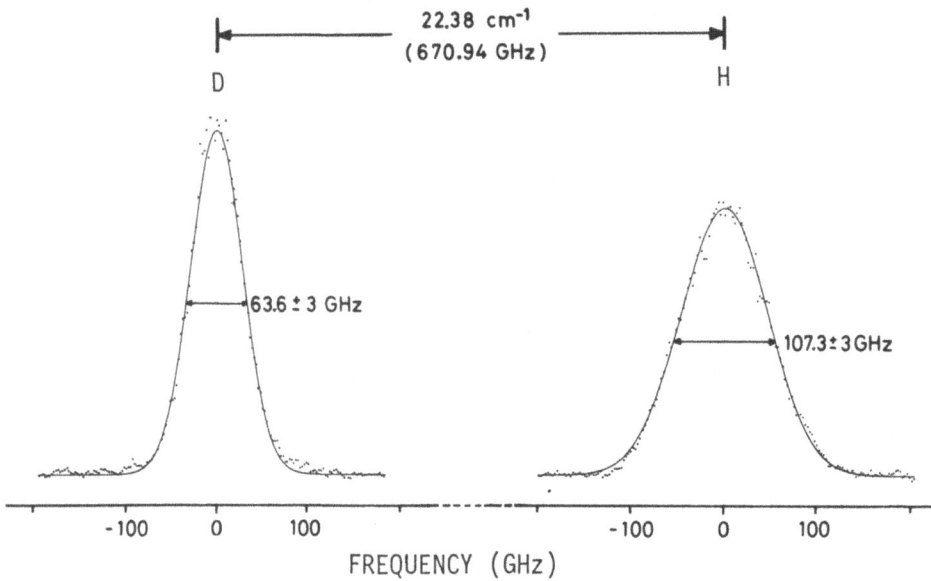

Fig. 3: Lyman-α line profiles of H and D atoms in the photodissociation of HDCO at 266 nm.

which in turn shows largely randomization of the energy in the
dissociation process. For RRKM-type decomposition the theoretical
ration of the yield of the two bimolecular channels, i. e. H +
DCO and D + HCO, should be 1.39 while the experimental value is
1.34 ± 0.3.

In evaluating the line width one has to consider that it is
determined, aside from the recoil energy of the atoms, also by
contributions from the Doppler width of the parent molecule velocity distribution and the laser bandwidth. The Doppler half-
width $\Delta \upsilon_P$ from the parent motion at 300 K is ~ 5.5 GHz. Folding
$\Delta \upsilon_P$ quadratically with the laser bandwidth of $\Delta \upsilon_L$ ~ 20 GHz ($\Delta \upsilon_0^2 =
\Delta \upsilon_L^2 + \Delta \upsilon_D^2$) yields an experimental bandwidth contribution of $\Delta \upsilon^0$ ~
21 GHz for atoms without recoil. Taking $\Delta \upsilon_0$ into account one ob-
tains the bandwidths $\Delta \upsilon_R$ due to the recoil of the atoms: (107 ±
3) GHz for H + DCO and (64 ± 3) GHz for D + HCO. Corresponding
to these bandwidths, the most probable velocities of H and D are
(7830 ± 240) m/s and (4640 ± 140) m/s, and the respective mean
kinetic energies ($\bar{E}_K = 3 m \tilde{v}^2/4$) are (3880 ± 240) cm^{-1} and (2720
± 170) cm^{-1}. Because of the fragment mass ratios the measured
atom velocities practically refer to the center-of-mass system of
the parent molecules. For the bimolecular decay channels one thus
would obtain the mean internal energies of the HCO and DCO frag-

ments as (5000 ± 180) cm^{-1} and (3900 ± 240) cm^{-1}, respectively, while the recoil energies of the radicals are (190 ± 10) cm^{-1} and (130 ± 10) cm^{-1}.

HYDROGEN MOLECULE

Two-photon-ionization

The absorption of the hydrogen molecule from the ground electronic state, $X^1\Sigma_g^+$, to the first excited state, $B^1\Sigma_u^+$, ranges for room temperature gas from $\sim$ 1108 to 1116 Å. State-selective one-photon excitation to the B-state by tunable vuv laser light, ω_{vuv}, followed by photoionization from this state with uv light, ω_{uv},

$$H_2(X^1\Sigma_g^+, v'' = 0, J'') + \hbar\omega_{vuv} \rightarrow H_2(B^1\Sigma_u^+, v', J') + \hbar\omega_{uv} \rightarrow H_2^+ + e$$

has recently been reported[12,13,14]. In our work[12,14], tunable vuv has been produced by sum frequency mixing ($\omega_{vuv} = 2\omega_{uv} + \omega_{vis}$) in krypton. Light in the visible, ω_{vis}, from a dye laser was doubled in a KDP crystal to obtain the uv component, ω_{uv}. Tuning ω_{vis} in the range 5540 - 5580 Å and ω_{uv} correspondingly in 2770 - 2790 Å yielded ω_{vuv} in the desired range 1108 - 1116 Å. The total photon energy $\hbar\omega_{vuv} + \hbar\omega_{uv}$ ranged thus from 126171-126545 cm^{-1}, exceeding the H$_2$ ionization potential IP$_0$(H$_2$) = 124417.2 cm^{-1} by 1754 to 2128 cm^{-1}.

Feasibility studies have been carried out with three different experimental set ups: (a) a simple ionization cell[14], (b) a quadrupole mass filter[14], and (c) a pulsed molecular beam system. Fig. 4b shows a spectrum obtained with the quadrupole mass spectrometer, where hydrogen was let into the ionization region through a capillary. The vuv and uv light beams of $\sim$ 1 mm and $\sim$ 3 mm diameter, respectively, passed through the ionization region perpendicular to the quadrupole axis. The spectrum shown in Fig. 4b has been taken at a hydrogen pressure of 2 x 10^{-4} Torr in the ionization region.

The ionization cell arrangement consisted essentially of a glass tube (1.8 cm diameter) with two parallel wires at $\sim$ 1.5 cm distance, serving as electrodes. The wires were located $\sim$ 3 cm behind a LiF lens through which the vuv and uv light beams entered the ionization cell. An example of a spectrum taken at an H$_2$ pressure of 3.4 x 10^{-3} is shown in fig. 4a.

As both experiments have been carried out with hydrogen in bulk the spectra (a) and (b) should show the same intensity distributions which, however, is not the case. They rather exhibit differences in two ways. Firstly, they differ from each other

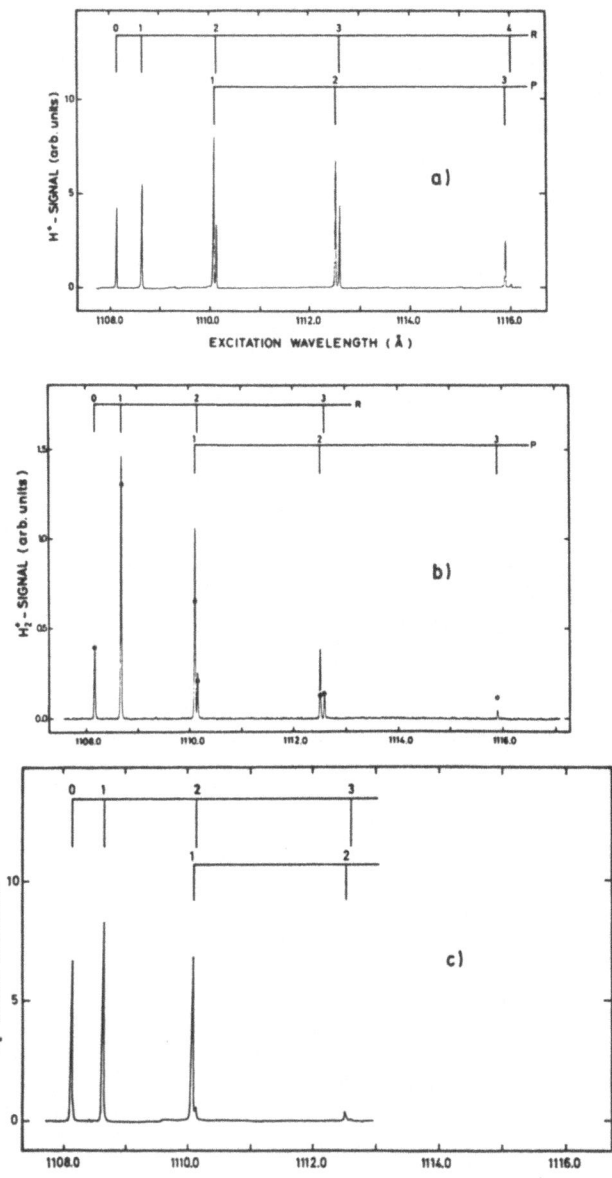

Fig. 4: State-selective two-photon excitation-ionization of H_2 molecules through the B-state. Experimental arrangements: (a) ionization cell, in bulk, (b) quadruple mass filter, in bulk, (c) crossed pulsed molecular beam with time-of-flight mass filter detection.

and, secondly, they deviate from the intensity distribution expected at room temperature indicated in spectrum (b) by the cir-
cles. A main cause for the deviation from the room temperature
distribution was given by the wavelength dependence of the uv and
vuv laser intensity, caused by the pressure dependence of the
phase matching condition. The krypton pressure was kept constant
at 190 Torr in this measurement, while the optimum phase matching
pressure varied from 190 Torr at 1115.9 Å to 340 Torr at 1108.1
Å. Differences between the spectra taken in the ionization cell
and in the mass filter are caused by preabsorption in the cell
experiments where higher H_2 pressures (2 x 10^{-5} to 4 x 10^{-2} Torr)
have been used[14]. As to the detection sensitivity, from the present experiments a state-selective detection sensitivity of
~10^{-10} Torr H_2 is derived for laser intensities now achievable in
the 1000 - 1200 Å region. More details of the work may be taken
from ref. 14.

Further experiments have been carried out in a two-stage molecular beam arrangement with a pulsed beam source of $\gtrsim$ 30 μsec
pulse duration. The molecular beam was crossed at right angle behind the scimmer orifice by the vuv and uv laser beams. The intersection was located 5.5 cm from the beam nozzle between the first
two parallel electrodes (1.2 cm apart) of a time-of-flight mass
spectrometer. The distance between scimmer orifice and beam
nozzle was 2.7 cm, so that the room temperature Doppler width of
the H_2 lines at 1110 Å wavelength was reduced to ~ 2 GHz in this
beam configuration. Vuv and uv has been generated in the same way
as in the bulk experiments before[14]. Fig. 4c shows an excitation-ionization spectrum taken at 5 bar stagnation pressure of
H_2. Comparison of the bulk and beam spectra shows that the intensity of the lines from rational levels J" > 1 are noticably
lower in the beam experiments, indicating cooling of the H_2 molecules in the expansion process. This is being investigated further.

Two-photon-spectroscopy

Excitation of the B-state makes two-photon spectroscopy and
investigation of photodissociation and photoionization from selectively prepared higher states possible,

$$H_2(X^1\Sigma_g^+, v', J') + \hbar\omega_{vuv} \rightarrow H_2(B^1\Sigma_u^+, v'', J'') + \hbar\omega_{uv} \rightarrow H_2^*$$

the two-photon absorption leads to final levels of H_2^* of equal
parity with the ground state, mostly still unidentified. We have
carried out feasibility experiments in the (0,0-band, H_2(X, v' =
0,J') $\rightarrow$ H_2(B,v" = 0,J"), and further excitation, H_2(B, v" = 0,
J") $\rightarrow$ H_2^* around the ionization limit.

a $H_2(X, v=0, J=1) \rightarrow H_2(B, v=0, J=2) \rightarrow H_2^+ + e$

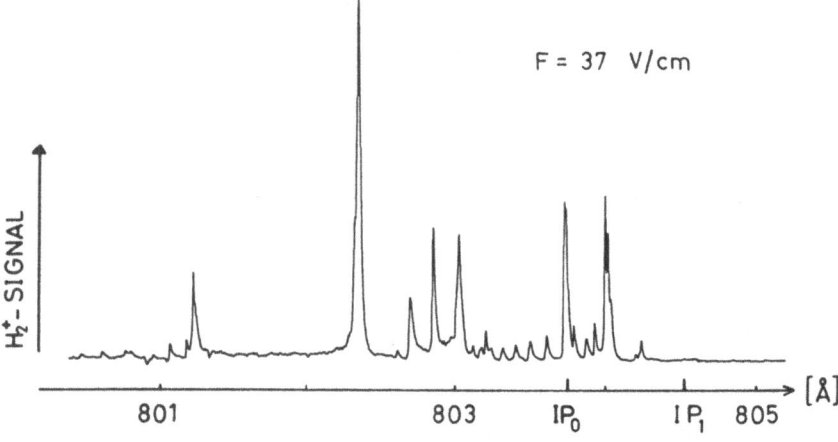

F = 37 V/cm

b $H_2(X, v=0, J=0) \rightarrow H_2(B, v=0, J=1) \rightarrow H_2^+ + e$

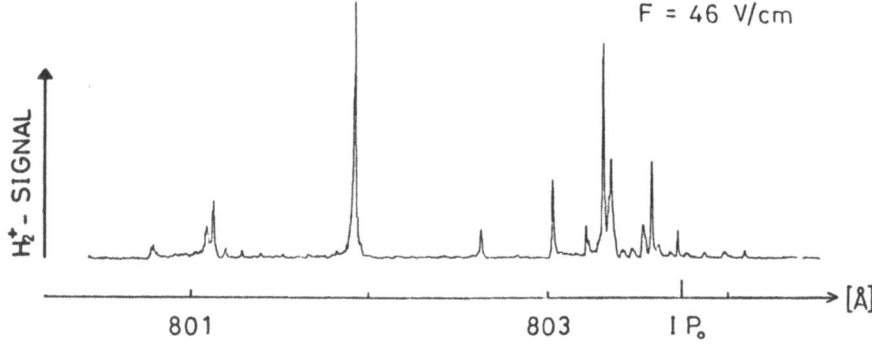

F = 46 V/cm

Fig. 5: H_2^+ production by state selective two-photon excitation: of H_2 molecule around the ionization limit with independently tunable vuv and uv light. Experimental arrangement crossed molecular beam. Intermediate levels: (a) (B, v = 0, J = 0); (b) (B, v = 0, J = 2).

For two-photon spectroscopy through the B-state interference from the longer-wavelength uv mixing light used to produce vuv must be excluded and a second uv laser source, independently tunable, must be employed. Different from the two-photon ionization

experiments, vuv radiation in the range 1108 - 1116 Å has therefore been produced by frequency tripling in krypton light in the
range 3324 - 3348 Å. Photo-ionization of H_2 molecules from the
B-state by this light did not occur. The tunable uv radiation,
ω_{uv}, for the second excitation step was obtained from an ex-
cimer-laser pumped dye system. The wavelength range of ω_{uv} was
chosen so that the total energy $h\omega_{vuv} + h\omega_{uv}$ of the two photons
was around the zero-field ionization potential of the H_2 molecule
$IP_o = 124417.2$ cm^{-1}, where IP_o refers to $H_2(X, v = 0, J = 0)$
$\rightarrow H_2^+(X, v = 0, J = 0)$. The experiments have been carried out in
the crossed molecular laser beam set-up mentioned in the pre-
ceeding section. Because of the short radiative lifetime of the
B-state ($\sim$ 0.5 nsec) both light pulses were synchronized such
that they arrived simultaneously at the molecular beam intersection.

Results are shown in Fig. 5 and 6 representing ion yields
from selected B-state levels (v = 0, J) when $h\omega_{uv}$ was scanned
through wavelength ranges including the ionization potential. In
taking the spectra Fig. 5a and b relatively weak electric fields
(37 and 46 V/cm) were present during excitation to extract the
ions into the mass filter, while the spectrum Fig. 6 has been obtained with zero field strength during excitation and a strong
field (2100 V/ cm) turned on after the laser pulses with a delay
time of $\sim$ 300 ns. The energy IP_o, indicated in the figures, refers to $H_2(X, v = 0, J = 0) - H_2^+(X, v = 0, J = 0)$, while IP_1 in
Fig. 5a refers to the energy of $H_2(X, v = 0, J = 1) - H_2^+(X, v =
0, J = 0)$.

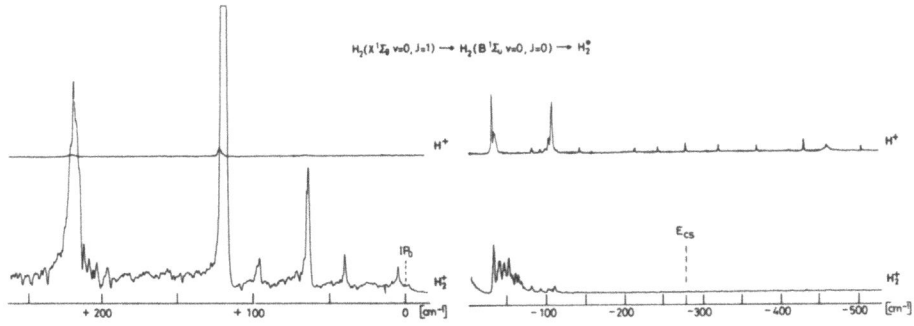

Fig. 6: Lower part: H_2^+ production by state selective two-
photon excitation of H_2 molecule around ionization limit,
field ionization at 2100 V/cm $\sim$ 300 nsec after excitation. E_{cs} classical field ionization saddle point energy.
Intermediate level (B, v = 0, J = 1).
Upper part: H^+ production (see text).

The spectra exhibit pronounced autoionization structures superimposed on a weak continuum like background. Possible states of the H_2^+ ion are determined by the energetics and selection rules. In the ionization spectrum from H_2(B, v = 0, J = 0) in Fig. 6 H_2^+(X, v = 0, J = 1) is the only product state possible. In this experiment field ionization is observed in addition to direct ionization, corresponding to 2100 V/cm field strength. The classical saddle point energy with respect to H_2^+(X, v = 0, J = 0) is E_{cs} = 280 cm^{-1}.

Aside from H_2^+ also H^+ is observed in the spectrum Fig. 6. The appearance of H^+ ions is explained by dissociation of H_2^* below the zero field ionization limit from high Rydberg states,

$$H_2(B, v = 0, J = 0) + \hbar\omega_{uv} \rightarrow H_2^*(ns, nd) \rightarrow H(n = 2) + H(ns = 1)$$

and subsequent photoionization of H(n = 2) by the uv light, similarly as in two-photon excitation-ionization of H atoms mentioned above. Field ionization of H_2^*(ns, nd) molecules does apparently not occur except in the range close to the zero field ionization limit. This may be explained by domination of the dissociation rate at lower energies.

ACKNOWLEDGEMENT

The experiments reported have been performed in collaboration with H. Rottke, H. Zacharias, R. Schmiedl and W. Meier.

REFERENCES

1   D.J. Brink, D. Proch, D. Basting, K. Hohla, P. Lokai,
     Laser und Optoelectronik 14, 41 (1982).
     H. Schomberg, H.F. Döbele and B. Rückle, Appl. Phys. B 30,
     131 (1983).

2   R. Wallenstein, Laser u. Optoelectronik 14, 29 (1982).

3   R. Wallenstein, Optics Comm. 33, 122 (1980).

4   H. Zacharias, H. Rottke, J. Danon and K.H. Welge,
     Optics Comm. 37, 15 (1981).

5   P. Bogen, R.W. Dreyfus, Y.T. Lie and H. Langer,
     J. Nuclear Materials 111 and 112, 75 (1982),
     P. Bogen, P. Martens, Verhandlg. Deutsche Phys. Ges. 18,
     444 (1983).

6   R. Hilbig and R. Wallenstein, to be published, IEEE J. Quantum
     Electronic (1983).

7   H. Zacharias, R. Schmiedl, R. Wallenstein and K.H. Welge,
     Laser 79, Opto-Electronics, ed. W. Waidelich, IPC Science
     and Technology Press, Gouldford (1979), pp. 74-80.

8   R. Wallenstein and H. Zacharias, Optics Comm. 36, 429 (1980).
9   H. Rottke, H. Zacharias and K.H. Welge, AIP Conference Pro-
    ceedings No. 90, "Laser Techniques for Extreme Ultraviolet
    Spectroscopy", Boulder 1982, ed. T.J. McIlrath, R.R. Freeman.
10 J.L. Kinsey, J. Chem. Phys. 66, 2560 (1977).
11 R. Schmiedl, H. Dugan, W. Meier and K.H. Welge,
    Z. Phys.-Atoms and Nuclei 304, 137 (1982).
12 K.H. Welge and H. Rottke, Conference Report Second European
    Workshop "Molecular Spectroscopy and Photon Induced Dynamics",
    De Eemhof, The Netherlands (1982).
13 E.E. Marinero, C.T. Rettner, R.N. Zare, A.H. Kung,
    Chem. Phys. Lett. 95, 486 (1983).
14 H. Rottke and K.H. Welge, in press, Chem. Phys. Lett.
    (1983).

LASER SPECTROSCOPY OF MOLECULAR IONS

Alan Carrington

Department of Chemistry
University of Southampton
Hampshire

INTRODUCTION

Important advances in the spectroscopy of molecular ions
during the past six years have arisen largely from the use of
lasers, both fixed and tunable in frequency. This review deals
with three of the most important areas, but it is by no means
comprehensive. In particular we do not deal with the Doppler-
limited electronic spectroscopy of molecular ions carried out
with pulsed tunable dye lasers, since this is discussed in the
articles by Leach [1].

We first describe the beautiful work initiated by Oka [2]
on the infrared spectra of ions, especially H_3^+, using a tunable
difference-frequency laser. The other two methods employ fixed-
frequency lasers, in which spectral scanning is achieved either by
means of tuning the molecular energy levels with a magnetic field,
or by employing the Doppler shift which arises when the laser
beam interacts with the rapidly moving ions in a high potential
ion beam.

INFRARED SPECTROSCOPY USING A DIFFERENCE-FREQUENCY LASER

Oka [2] has observed the infrared absorption spectrum of
the H_3^+ ion using the difference-frequency laser system originally
developed by Pine [3]. Radiation from an argon ion laser and
from a tunable dye laser are mixed in a lithium niobate crystal
to produce infrared radiation of a few microwatts, tunable over
the range 4400 to 2400 cm^{-1}. The infrared beam was passed through
a multiple reflection cell containing a hydrogen discharge plasma.

117

The effective path length was 32 m and frequency modulation of
the argon laser allowed phase-sensitive detection of the infrared
beam after transmission. More than 20 lines of H_3^+ have now been
observed, arising form rotational components of the fundamental
degenerate vibrational stretching mode. Theoretical calculations
show that H_3^+ is expected to exhibit equilateral triangular
geometry, and calculations by Carney and Porter [4] of the
vibration-rotation spectrum are in good agreement with Oka's
measurements. Oka is now in the process of building a new
spectrometer system with increased sensitivity: he has also
succeeded in observing the H_3^+ spectrum using an infrared tunable
diode laser system [5].

Oka's initial spectrometer system has since been used by
Bernath and Amano [6] to study the fundamental vibrational band
of HeH^+, and by Wong, Bernath and Amano [7] to investigate the
fundamental vibrational band of $^{20}NeH^+$ and $^{22}NeH^+$. Accurate
molecular constants have been obtained for both of these ions.

The infrared emission spectrum of ArH^+ has also been
observed by Brault and Davis [8] using a Fourier transform
spectrometer. They have measured many rotational components of
five different vibration-rotation bands. It is now clear that
for ions which are the major species in discharge plasmas, further
absorption and emmision studies are likely to be reported.

FAR INFRARED LASER MAGNETIC RESONANCE

Far infrared laser magnetic resonance has proved to be an
extremely sensitive technique for detecting rotational transitions
in open shell molecules, where the paramagnetism arising from
electrons of unpaired spin means that the molecular energy levels
can usually be tuned with an applied magnetic field. The
possibility of using this method to detect the spectra of molecular
ions in discharge plasmas has long been apparent, and successful
experiments have now been reported for HBr^+ by Saykally and
Evenson [9] and for HCl^+ by Ray, Lubic and Saykally [10]. The
far infrared laser radiation is produced by excitation of an
appropiate gas (for example, CH_3OH) with infrared radiation from
a carbon dioxide laser in a suitable gain cell. Part of the laser
cavity is isolated from the laser plasma with a suitable membrane
and serves as a discharge cell in which the molecular ions are
produced. This section of the laser cavity is placed either
between the pole faces of an electromagnet, or inside the bore
of a solenoid. A sample of the far infrared radiation is detected
with a suitable low temperature bolometer. Auxiliary magnetic
field modulation allows the use of phase-sensitive detection.
For both HBr^+ and HCl^+, which have $^2\Pi$ ground states, a number of
rotational transitions were detected and since the Doppler width

at far infrared wavelengths is comparatively small, nuclear
hyperfine structure from the halogen isotopes was observed,
yielding values of the magnetic and electric quadrupole coupling
constants.

ELECTRONIC AND VIBRATIONAL SPECTROSCOPY OF MOLECULAR ION BEAMS

The use of molecular ion beams as a medium for spectroscopy
is attractive because the technology of mass spectrometry is
highly developed, and a great many ions of importance have been
identified in the mass spectrometer. In general, however, the
ion density even in a strong beam (i.e. 10^{-7} A or 10^{12} ions/sec)
is sufficiently low that indirect methods of detection are usually
required. An important exception has been described recently by
Rosner, Gaily and Holt [11] who have recorded the laser-induced
fluorescence spectrum of N_2^+ in an ion beam. By aligning the laser
beam (wavelength 428 nm) to be collinear with the ion beam they
are able to exploit two advantageous features of kilovolt ion
beams, as discussed in detail by Carrington [12]. Firstly,
sweeping the potential of the ion beam yields a frequency scan
via the Doppler shift. Secondly, the effective Doppler width
is much reduced by kinematic compression or velocity bunching
effects. Consequently line widths down to 60 MHz were observed,
enabling resolution of the ^{14}N hyperfine structure in the
electronic spectrum.

The first high-resolution studies of an infrared spectrum
were reported by Wing, Ruff, Lamb and Spezeski [13] who detected
vibration-rotation transitions of the HD^+ ion. They used a
carbon monoxide laser as the radiation source and employed Doppler
tuning. They detected the infrared resonances by a novel charge-
exchange scheme. Introduction of neutral H_2 gas attenuates the
HD^+ beam because of charge-exchange; vibrational excitation of
the HD^+ ions by the infrared laser results in a change in the
charge-exhange cross-section, and consequently a change in the
HD^+ beam intensity. Consequently the vibration-rotation spectrum
is recorded by scanning the ion beam potential and monitoring
changes in the HD^+ beam intensity. Wing et al were able to
measure a number of vibration-rotation energies for the v=0 to
3 levels, thus providing stringent experimental tests of the many
theoretical calculations which have been performed for this
simple one-electron moecule. Because of the kinematic compression
effect mentioned earlier line widths of only a few MHz were
observed, so that proton hyperfine structure could be resolved.
More recently Tolliver, Kyrara and Wing [14] have used similar
charge-exchange detection to study, for the first time, the
vibration-rotation spectrum of HeH^+, and simultaneously with
Oka's work on H_3^+, vibration-rotation transitions in D_3^+ have been
detected by Shy, Farley, Lamb and Wing [15], and in H_2D^+ by Shy,
Farley and Wing [16]. Charge-exchange detection can also be used

to study electronic spectra, as has been demonstrated by Carrington, Milverton and Sarre [17] for the first electronic band system of CO^+. They were able to resolve ^{13}C nuclear hyperfine structure using a $^{13}CO^+$ beam, and obtain hyperfine constants for both the ground and first excited electronic states of CO^+.

The most sensitive experiments on molecular ion beams are those which depend upon predissociation. The first example was that of O_2^+ in which laser excitation from the $^4\Pi_u$ to the $^4\Sigma_g^-$ state is detected by predissociation of the upper state to yield O^+ ions which are easily separated from the parent O_2^+ ions [18]. Since the O^+ photofragment ions can be detected with essentially 100% efficiency, the predissociation technique has the sensitivity advantages associated with emission spectroscopy, even though an absorption spectrum is recorded. Similar studies of the electronic spectrum of the CH^+ ion have been described recently, the spectrum being detected by monitoring the C^+ photofragments [19]. The electronic predissociation spectrum of H_2S^+ has also been described by Edwards, Maclean and Sarre [20], and, very recently, that of the PH_2^+ ion [21]. Elsewhere in this volume [22] predissociation spectra obtained with infrared radiation by Carrington and his colleagues are discussed in detail.

I wish to thank the Royal Society for a Research Professorship and the Science and Engineering Research Council for their financial support.

REFERENCES

[1] S. Leach, this volume (1982).
[2] T.Oka, Phys.Rev.Lett.,45,531 (1980).
[3] A.S.Pine, J.Opt.Soc.Amer.,64,1683 (1974); 66,97 (1976).
[4] G.D.Carney and R.N.Porter, J.Chem.Phys.,65,3547 (1976).
[5] T.Oka, private communication.
[6] P.Bernath and T.Amano, Phys.Rev.Lett.,48,20 (1982).
[7] M.Wong, P.Bernath and T.Amano, to be published.
[8] J.W.Brault and S.P.Davis, Physica Scripta,25,268 (1982).
[9] R.F.Saykally and K.M.Evenson, Phys.Rev.Lett.,43,515 (1979).
[10] D.Ray, K.G.Lubic and T.J.Saykally, Mol.Phys.,46,217 (1982).
[11] S.D.Rosner, T.D.Gaily and R.A.Holt, to be published.
[12] A.Carrington, Proc.Roy.Soc.,A367,433 (1979).
[13] W.H.Wing, G.A.Ruff, W.E.Lamb and J.J.Spezeski, Phys.Rev.Lett., 36,1488 (1976).
[14] D.E.Tolliver, G.A.Kyrala and W.H.Wing, Phys.Rev.Lett., 43,1719 (1979).
[15] J.-T.Shy, J.W.Farley, W.E.Lamb and W.H.Wing, Phys.Rev.Lett., 45,535 (1980).
[16] J.-T.Shy, J.W.Farley and W.H.Wing, Phys.Rev.,A24,1146 (1981).

[17] A.Carrington, D.R.J.Milverton and P.J.Sarre, Mol.Phys.,
     35,1505 (1978).
[18] A.Carrington, P.G.Roberts and P.J.Sarre, Mol.Phys.,35,1523
     (1978);

     J.T.Moseley, M.Tadjeddine, J.Durup, J.B.Ozenne, C.Pernot
     and A.Tabche-Fouhaille, Phys.Rev.Lett.,37,891 (1976).
[19] A.Carrington and P.J.sarre, J.Phys.,Paris,40,C1 (1979);
     P.C.Cosby, H.Helm and J.T.Moseley, Astrophys.J.,235,52 (1980);
     M.M.Graff and J.T.Moseley, Chem.Phys.Lett.,83,97 (1981).
[20] C.P.Edwards, C.S.Maclean and P.J.Sarre, Chem.Phys.Lett.,
     87,11 (1982).
[21] P.J.Sarre, private communication.
[22] A.Carrington, this volume (1982).

PHOTODISSOCIATION DYNAMICS EXPERIMENTS WITH NO_2

Karl H. Welge

Fakultät f. Physik
Universität Bielefeld
4800 Bielefeld, Fed. Rep. of Germany

INTRODUCTION

Stimulated by the progress in laser techniques the experimental, and by that also the theoretical investigation of photodissociation processes has gained considerable interest in recent years. For small polyatomic molecules, particularly triatomics, theory has advanced to the point where detailed dissociation cross sections on quantum state level can be treated. Using laser both for excitation of the dissociation process and for observation of products experiments of state-to-state type are now feasible in principle, though the complexity of the problem with the various dynamical parameters has confined studies to one or the other experimental aspect. Previous experiments have thus been concerned with internal product energy analyses, mostly by laser induced fluorescence, or with kinetic energy and angular distribution measurement by time-of-flight techniques. Also, experiments have been conducted so far almost exclusively at fixed dissociation energies, leaving the fundamentally important question of the spectral dependence very much open.

In this paper we report experiments carried out in our laboratory on the one-photon photodissociation of the NO_2 molecule,

$$NO_2(X^2A_1; \nu_i; N_{K_a K_c}) + h\nu \rightarrow NO(X^2\Pi_{1/2,3/2}; v, J) + O(^3P_{2,1,0})$$

In previous experiments with the NO_2 parent gas at room temperature and in bulk we have investigated the internal energy distribution (vibration and rotation) of the NO fragments at fixed dissociation wavelengths, namely 337 (1), 308 and 351 nm (2).

Since the internal energy of NO_2 at 300 K (practically only rotational) is $\sim$ 310 cm^{-1} which is appreciable compared with the excess energy we have carried out experiments in adiabatically cooled seeded beams so that the molecules are contained in a few lowest rotational states. Working with internally cold parent molecules is of course desirable with respect to the energetics of the whole dissociation process. The definition of the initial quantum state of the system is, however, even more essential with respect to detailed dynamics of the excitation and decay mechanism. Quantum state specific initial state preparation and product analysis is particularly important in cases of predissociation with sharply structured spectra like in NO_2. State-to-state experiments will provide information about the "channel specific" partial absorption cross section, and by that deeper insight in the spectroscopy and molecular dynamics of such complex systems like NO_2 with its very strong interaction between excited and ground electronic states.

Because predissociation occurs already at the dissociation energy, $D_o = (25130 + 1)$ cm^{-1} (3) we have performed the experiments in the immediate vicinity of the dissociation limit. Different from previous studies, we have furthermore observed detailed dissociation rates as function of the dissociation photon energy for single, well defined product quantum state channels. In this paper we provide a brief summary of the experiments and of some representative results. More details will be given in a forthcoming publication.

EXPERIMENTAL

The experimental set up consisted in its main parts of the reaction chamber with a pulsed inlet nozzle valve ($\sim$ 1 ms pulse duration) pumped by a booster pump, the fluorescence detection system, and tunable pulsed dissociation and probe lasers. The laser beams crossed each other at 60° angle on the molecular beam axis at $\sim$ 10 mm distance from the valve orifice. The essential region of the beam intersection was determined by the laser beam diameters of $\sim$ 2 mm and $\sim$ 1 mm, respectively, at the intersection. Fluorescence excited by the laser was received through an f/1, collimating lens system by a photomultiplier (RCA 1P28). The lasers ($\sim$ 7 ns pulse duration, 10 Hz repetition rate), were operated such that the dissociation laser was fired $\sim$ 300 μs after the beginning of the molecular beam pulse and the probe laser pulse was delayed by 50 ns against the dissociation pulse. The photomultiplier signal was received through an electronic gate of 0.5 μs width, opened 50 ns after the probe laser pulse. The gate width was compatible with the radiative lifetime ($\sim$ 0.2 μs) of the $NO(A^2\Sigma)$ state.

Probing of the NO fragments has been performed by laser induced fluorescence excitation in the (0,0) band of the NO(X-A) transition

in the wavelength region around 2350 Å. The dissociation has been
investigated in an energy region a few hundred wave-numbers wide
around the dissociation limit, D_0 = 25130 cm^{-1} ($\hat{=}$ 3979 Å).

INITIAL STATE POPULATION OF NO_2

Nitrogen dioxid was cooled by adiabatic expansion seeded in
helium. Cooling was monitored by LIF analysis of NO_2 in the spectral
region of the $(\tilde{X}^2A_1 - \tilde{A}^2B_2)$ transition around 15436.0 cm^{-1}. The
degree of cooling depended on the helium stagnation pressure and
in particular also on the NO_2 partial pressure. For instance, at
10 bar He with 0.6 % NO_2 the population was contained in the NO_2
rotational levels, 0_{00} and, 2_{02} and 4_{04} with a distribution ratio
of 45 : 52 : 3 %, as indicated by the height of the bars in the
middle part of Fig. 1. The other two distributions with population

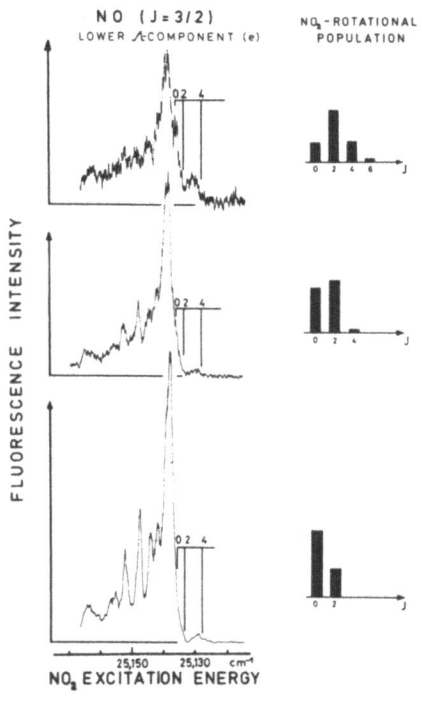

Fig. 1: Left side: Spectral dependence of the dissocia-
tion yield of $NO(\Pi_{1/2}$;
0, J = 3/2) + $O(^3P_2)$.
Right side: Corresponding
initial state distribu-
tions of NO_2.

ratios 21 : 55 : 22 : 3 % and 70 : 30 % in the levels indicated in
the figure refer to 10 bar He with 1.6 % NO_2 and 10 bar He with
0.15 % NO_2, respectively. Although the population distributions,
deviate from thermal equilibrium they correspond within a rough
approximation to rotational NO_2 temperatures of 5.5, 2.5 and 1.8 K,
respectively.

ROTATIONAL POPULATION DISTRIBUTION OF NO

In this work we have carried out experiments in a rather narrow range of excess energy ($E_{exc} = E(h\nu) - D_o$), namely $E_{exc} \lesssim$ 200 cm^{-1}. In this interval dissociation channels are limited to rather few NO rotational levels and correspondingly low product recoil energies.

Fig. 2 shows two examples of NO rotational distributions (plotted semilogarithmically) measured at excess energies $E_{exc} = 32$ and 122 cm^{-1} and NO$_2$ rotational temperatures of 7 and 23 K, respectively. The curves start with relatively steep slopes at low rotational energies and tend to flatten at higher energies. Points beyond the excess energy limits result from the residual internal NO$_2$ energy at the respective temperatures.

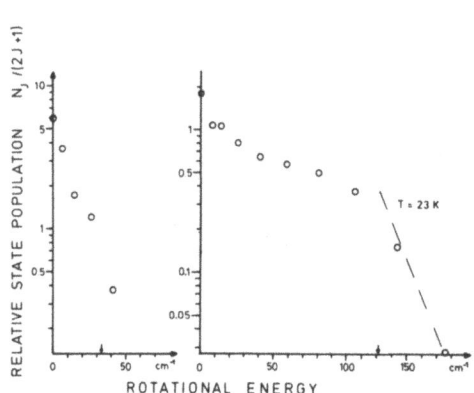

Fig. 2: Rotational energy distribution of NO at excess energies of 32 and 122 cm^{-1}, plotted in semi-logarithmic form. NO$_2$ rotational "temperatures" 7 and 23 K, respectively.

This type rotational NO population dependence with low rotational "temperature" NO fragments in low levels to high temperature fragments in higher levels has been observed also in the previous experiments with NO$_2$ at room temperature and with much higher photon excess energies (1,2). It apparently is a general behaviour of the NO$_2$ photodissociation process. Aside from the qualitative explanation of rotationally hot diatomic fragments occurring in processes where the ground and excited states have different equilibrium bond angles, a more quantitative theoretical treatment of the observed distributions is not available. Clearly, the smooth variation from low to high temperature fragment distribution speaks against a mechanism that would involve one or the other discret different reaction path resulting in molecular fragments with correspondingly discernable internal state "temperatures".

CHANNEL SPECIFIC DISSOCIATION YIELD

At excess energies $E_{exc} \leq 127$ cm^{-1} and for cold NO$_2$ only decay channels

NO($\Pi_{1/2}$; 0, J $\leq$ 15/2) + O($^3P_2$)

are energetically possible (Fig. 3). Tuning the probe laser to lines from levels J $\leq$ 15/2 in the $\Pi_{1/2}$ manifold and scanning the dissociation laser wavelength will thus allow to measure the spectral dependence of the dissociation yield (or cross section) for channels where both products are in single defined states. The same is true

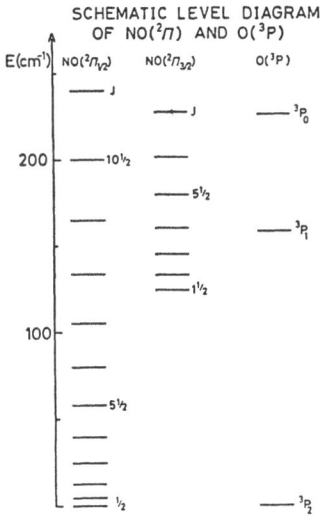

Fig. 3: Energy diagramme of NO and O internal states.

for the three lowest states J $\leq$ 7/2 of the $\Pi_{3/2}$ manifold (Fig. 3), that is for channels

NO($\Pi_{3/2}$; 0, J $\leq$ 7/2) + O($^3P_2$).

Furthermore, excitation of lines in different branches is selective with respect to Λ doubling components of a rotational level. In the present case lower Λ components "e" belong to the branches (R$_{11}$ + Q$_{21}$) and P$_{11}$ and the upper ones "f" to (Q$_{11}$ + P$_{21}$) and R$_{21}$.

Fig. 4 shows light traced the yield of the energetically lowest decay channel with NO($\Pi_{1/2}$; 0, J = 1/2) and the lower Λ component. The heavier traced curve represents the LIF excitation spectrum of NO$_2$. Both measurements have been made with NO$_2$ at 2.5 K temperature. The dashed-dotted curve shows for comparison the absorption spectrum of NO$_2$ in bulk at 235 K taken from Bass et al. (4). The energies of the NO$_2$ rotational levels 0_{00}, 2_{02} and 4_{04} are indicated in the figure with the 0_{00} level at the D$_0$ limit. The rotational population distribution in the levels has been 45 : 52 : 3 % in this experiment.

The onset of the dissociation coincides closely with the dis-
appearance of the NO_2 fluorescence. Because the energetics of the
dissociation was rather well defined, the experiment provides an
absolute determination of $D_0(O - NO)$, with an accuracy given by the
parent state population distribution and the laser band width. We
obtain $D_0 = (25130.6 \pm 0.6)$ cm^{-1} as compared to $D_0 = (25129.9 \pm 1.2)$ cm^{-1} derived in a previous work (3) where, however, the $\overline{NO}_2$
fluorescence excitation spectrum has been observed only.

The onset of the dissociation into the lowest channel does not
occur at the dissociation energy D_0 but already at energies lower
by that of the 2_{02} and 4_{04} states. This shows that the parent molecule rotation is efficiently coupled to the decay mechanism.

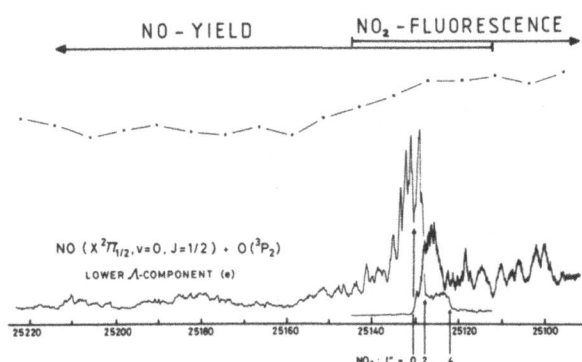

Fig. 4:
Light trace: Spectral
yield dependence of
product channel $NO(\Pi_{1/2}$;
0, J = 1/2) and $O(^3P_2)$
Heavy trace: NO_2 laser
excited fluorescence
spectrum.
Dashed-dotted trace:
NO_2 absorption at 235 K.

The dissociation dynamics depends of course principally on the
quantum state structure of the system. The dependence is demonstrated
by comparing the bulk absorption with the detailed, channel specific
dissociation spectrum in Fig. 4. The pronounced structure observed
in state selected experiments is naturally washed out by superposition of many detailed reaction channels occurring simultaneously
at higher temperatures.

The NO_2 dissociation apparently is particularly sensitive as
to the initial parent molecule state distribution. This is shown
by an experiment (Fig. 1) where the spectral yield of the channel
$NO(\Pi_{1/2}$; 0, J = 3/2) + $O(^3P_2)$ has been investigated with NO_2 cooled
to different temperatures, i. e. initial state distributions. As
can be seen the yield spectra exhibit very pronounced structural
differences in detail when the population is shifted in the two
lowest NO_2 rotational levels 0_{00} and 2_{02}, with an energy difference
of only 2 cm^{-1}. Experiments with pure state preparation in 0_{00} only
are obviously necessary to fully identify the channel specific
spectral dissociation structure. In this case homogeneous line

widths could be observed yielding direct information about the decay
lifetime with respect to the particular reaction channel.

Fig. 5 shows the spectral yield of four dissociation channels
with $NO(^2\Pi_{1/2})$ in $J = 1/2$, $3/2$, $5/2$ and $7/2$, and the lower Λ-compo-
nent "e". The NO_2 rotational temperature has been 2.5 K in the experiment. With increasing internal energy of NO the signal intensity
decreases because the number of channels gets larger and the total
absorption accordingly is distributed among more reaction routes.

A common behaviour is observed, namely the sharp onset of each
decay channel at its respective energy limit and the decrease of
the yield as next channel becomes possible. This indicates that rotational excitation is the preferential degree of freedom for the
excess energy.

$$NO\ (X^2\Pi_{1/2}, v=0, J\ , e)\ +\ O\ (^3P_2)$$

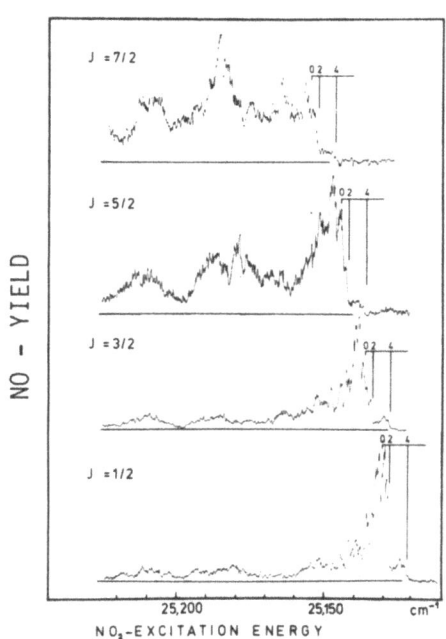

Fig. 5:
Spectral yield dependence
of product channels with
$NO(\Pi_{1/2}$, v=0) in J=1/2,
3/2, 5/2 and 7/2 and
$O(^3P_2)$. NO_2 rotational
temperature 2.5 K.

Another feature is that the dissociation yield of individual
channels exhibit sharp structures, particularly in the vicinity of
the threshold energy, indicating the predissociative nature of the
process. As mentioned above, more structure is expected with NO_2
prepared in the lowest level only.

Λ-COMPONENT SPECIFIC DISTRIBUTION

The lower and upper Λ components "e" and "f" are distinguished through the selection rules by transitions in the different branches as mentioned above. Since the Λ type splitting depends on symmetry properties of the states involved information about the dissociation mechanism may eventually be derived from Λ level distributions.

Fig. 6 shows a first result of experiments carried out in this respect. It compares the spectral yield dependence of the $NO(\Pi_{1/2}$, 0, J = 1/2) + $O(^3P_2)$ channel for the two Λ components. The initial NO_2 rotational state distribution was 45 : 52 : 3 % in the levels 0_{00}, 2_{02}, 4_{04}.

$NO\ (X\,^2\Pi_{1/2}, v=0, J=1/2) + O\,(^3P_2)$

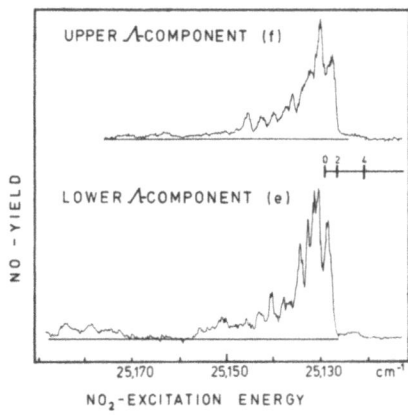

Fig. 6: Spectral yield dependence of $NO(\Pi_{1/2}; J=1/2)$ + $O(^3P_2)$ for Λ components "e" and "f" of J=1/2.

As can be seen the two Λ specific channels exhibit in details essentially different structures. Current experiments show that this different behaviour occurs also in decay channels with higher rotational levels of the NO fragment.

Again, rotational cooling of NO_2 into 0_{00} alone is necessary in order to observe the distribution on a pure state-to-state level.

ACKNOWLEDGEMENT

This work has been performed by U. Robra and is part of his Ph. D. thesis (5).

LITERATURE

1. H. Zacharias, M. Geilhaupt, K. Meier and K. H. Welge; J. Chem.
    Phys. 74, 218 (1981).

2. H. Zacharias, K. Meier and K. H. Welge; in "Energy Storage and
    Redistribution in Molecules" edit. J. Hinze, Plenum N. Y.
    (1983).

3. C. H. Chen, D. W. Clark, M. G. Payne and S. D. Kramer; Optics
    Comm. 32, 391 (1980).

4. A. M. Bass, A. E. Ledford, Jr., and A. H. Laufer; J. Res.
    Natl. Bur. Stand. 80A, 143 (1976).

5. U. Robra; Ph. D. Thesis, Univ. Bielefeld (1983).

COLLISION INDUCED MODE SELECTIVE ENERGY TRANSFER

IN METHYLFLUORIDE

R. Stender and J. Wolfrum

Max Planck-Institut für Strömungsforschung
D-3400 Göttingen, Böttingerstraße 4-8
and
Physikalisch-Chemisches Institut der
Ruprecht-Karls-Universität Heidelberg
D-6900 Heidelberg, Im Neuenheimer Feld 253

ABSTRACT

After single quantum excitation of the C-F stretching vibration (ν_3) in CH_3F with a CO_2 laser pathways for collision induced mode selective energy transfer into C-H bending and stretching modes were observed by laser induced infrared fluorescence. Until now, all the vibrational states of CH_3F were assumed to come into V-V equilibrium with the pumped ν_3 levels on the same time scale. The experiments reported here show that selective pathways with quite different time scales exist. In the case of pure CH_3F, vibrational energy is transferred first via "up the ladder" processes in the pumped ν_3 mode. Local coriolis resonances between $3\nu_3$ and ν_4 at high J, K states can explain the fast, but not very efficient process

$$CH_3F(3\nu_3) + CH_3F \rightleftharpoons CH_3F(\nu_4) + CH_3F$$

which is suggested to be responsible for a partial filling of the C-H stretch vibrations (ν_4). In a mixture of CH_3F and rare gas atoms the intermode energy gaps between ν_3, ν_6 and ν_2, ν_5 are surmounted in collisions of the excited $CH_3F(\nu_3)$ molecule with rare gas atoms. Because ν_1 and $2\nu_5$ are mixed by Fermi resonance, the V-V process leads directly to an efficient filling of the C-H stretch vibration (ν_1)

$$2CH_3F(\nu_5) \rightleftharpoons CH_3F(2\nu_5) + CH_3F$$

INTRODUCTION

Laser light sources offer a number of new ways for a controlled transfer of external energy into chemically reacting molecules. The high quantum flux, monochromaticity, coherence and short pulse duration of infrared lasers can be used to obtain new insights into the microscopic dynamic of chemical reaction by selective preparation of vibrationally excited molecules in the electronic ground state. This knowledge is also of basic interest for potential practical applications, in which external energy is used to drive chemical reactions more selectively than by using thermal heating. In this respect it is an especially intriguing question how the chemical reactivity of a polyatomic molecule changes if the same total energy is localized in different ways in certain vibrational modes. Recent investigations on unimolecular decompositions of highly vibrationally excited molecules have shown that the intramolecular energy transfer between the excited and other modes is extremely fast. This observation was made in experiments using infrared multiphoton excitation as well as in excitation of overtone vibrations[1,2] and ultraviolet eletronic excitation followed by rapid internal conversion[3]. In contrast the time scale for mode selective vibrational excitation of polyatomic molecules becomes much longer, if energetically lower levels far below the dissociation threshold are excited. The energy transfer here is governed by the limited number of pathways for the flow of energy between the discrete vibrational modes. On the other hand the potential energy barrier for the chemical reaction can be made comparable to the energy of one or few vibrational quanta if one turns from unimolecular to bimolecular reactions with a suitable reaction barrier.

In order to obtain information about the bimolecular chemical reaction of a polyatomic molecule which is excited in a specific vibrational mode one must first try to decouple the vibrational energy exchange from the removal of the excited molecule by interaction with the added reactant: Using a low partial pressure of the excited molecules and spectral and time resolved detection of the infrared emission from the different excited modes a direct observation of the energy flow is possible. For such investigations substituted methanes are well suited molecules. A great deal of information concerning vibration to vibration transfer has been obtained for the molecule CH_3F mainly due to the pioneering studies of Flynn and coworkers[4] using laser induced infrared fluorescence, infrared double resonance and time resolved thermal lensing experiments.

EXPERIMENTAL

A higher detection sensitivity for the laser induced fluorescence could be achieved in our experiments by a Welsh type mirror system for the collection of the infrared photons and two plane mirrors for multiple reflection of the exciting CO_2-laser beam (see Fig. 1). This gave an improvement of a factor 25 compared to fluorescence cells without collecting mirrors. The TEA CO_2-laser used is grating tuned to the P(20) line at 1048,85 cm^{-1} and excites the ν_3 C-F stretching mode of CH_3F. Cooled LiF and MgF_2 windows, several interference filters, GeHg (4K) and InSb (77K) detectors were used for time and spectral resolved fluorescence monitoring. The signal to noise ratio is improved by a transient recorder combined with a signal averaging system.

With the CO_2-laser pulses used (200 ns, 500 mJ) a fraction of 0.4 from the CH_3F molecules is excited to the ν_3 (v=1) level. The activation of several vibrational modes in CH_3F was studied by measuring the rise time of the fluorescence in the levels $2\nu_3$ (overtone of the pumped ν_3 level) at 2083 cm^{-1}, ν_6 (C-H bend) at 1181 cm^{-1}, ν_2, ν_5 (C-H bend) at 1460 cm^{-1} and ν_1, ν_4 (C-H stretch) around 3000 cm^{-1}.

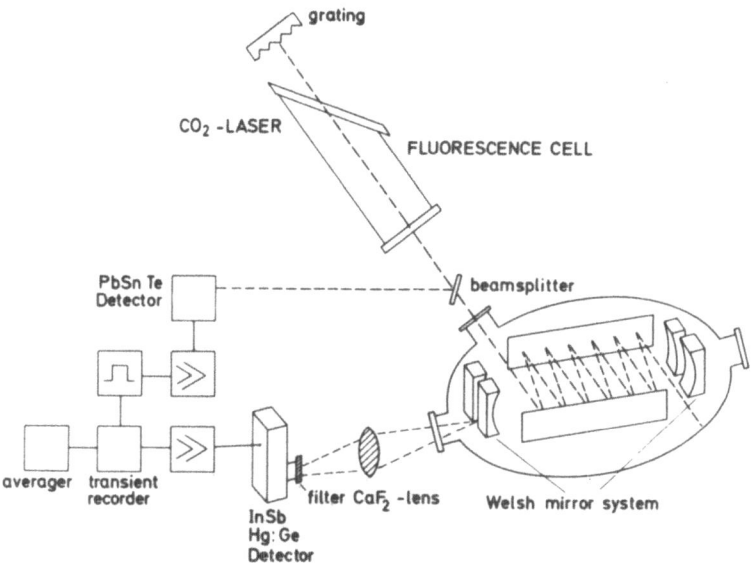

Fig. 1.   Schematic of the experimental apparatus for sensitive CO_2-laser induced infrared fluorescence detection

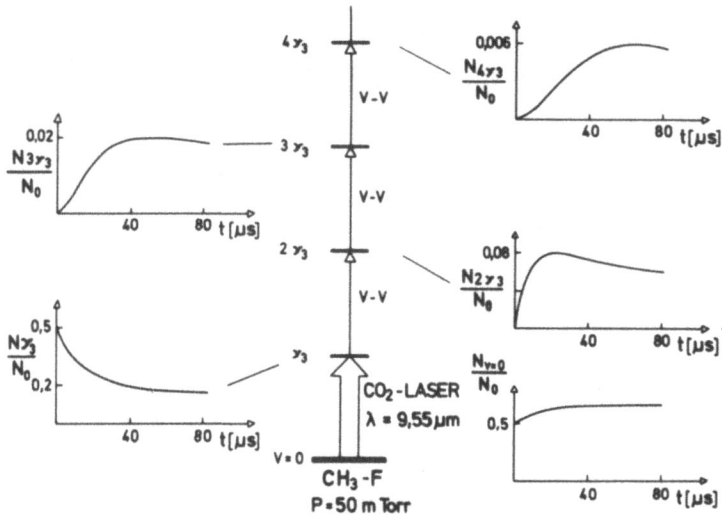

Fig. 2.   Measured and calculated time profiles of the ν_3 levels

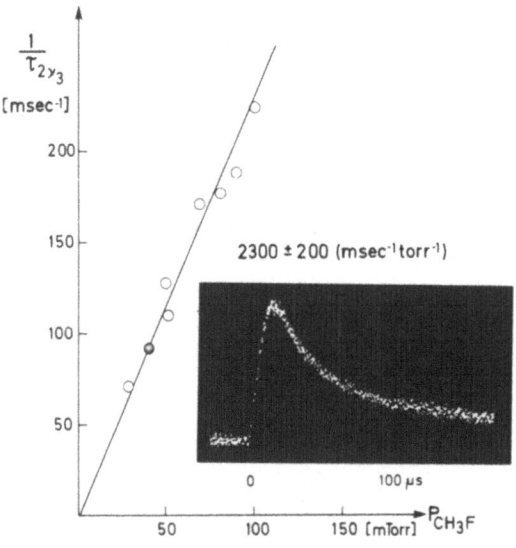

$CH_3F\,(\nu_3)\;+\;CH_3F\,(\nu_3)\rightleftharpoons CH_3F\,(2\nu_3)\;+\;CH_3F$

Fig. 3.   Measured time profile of the $2\nu_3$ level

Fig. 2 shows the measured and calculated time resolved population profiles of the ν_3 levels after laser excitation of the ν_3 mode at a CH_3F pressure of 50 mtorr. The experimental curves are the superposition of a rising and a decaying part. The mean collision time of CH_3F molecules at 50 mtorr is 2.4 µsec, which is more than ten times the laser pulse width so that nearly collision free conditions are achieved during absorption of CO_2-photons. The absorption process itself is a single photon event as indicated by the well resolved fluorescence rise of the $2\nu_3$ level (see Fig. 3).

A nearly ten times risetime is observed for the ν_2, ν_5-levels under these experimental conditions. But, as a very remarkable result, the rise time of the energetically much higher located-ν_1, ν_4-levels is faster.

Linear least-squares-fits, performed on the risetime data of the observed vibrations for pressures up to 250 mtorr are given in Table I. Based on the laser induced fluorescence studies a mechanism for the vibrational energy transfer from the pumped ν_3 level to the ν_1, ν_4 mode can be deducted. The overtone $2\nu_3$ is populated by the nearly resonant V-V process

$$CH_3F(\nu_3) + CH_3F(\nu_3) \rightleftharpoons H_3F(2\nu_3) + CH_3F + 16 \text{ cm}^{-1} \qquad (1)$$

in about 3.7 gaskinetic collisions. The cross section reported

Table I

Level Energy	Rate of Rise	gaskinetic collisions
$2\nu_3$ 2082 cm^{-1}	(2300 ± 200) msec^{-1} torr^{-1}	3.7
ν_2, ν_5 1460 cm^{-1}	(255 ± 15) msec^{-1} torr^{-1}	38
ν_1, ν'_4 3000 cm^{-1}	(920 ± 80) msec^{-1} torr^{-1}	9.2

by Earl et al.[15] for the "up-the-ladder" process (1) was 1/7 gas kinetic. In a recent study by Sheorey and Flynn[4] 3.3 gas kinetic collisions are reported.

The ν_3-mode is further populated by ladder climbing processes such as

$$CH_3F(2\nu_3) + CH_3F(\nu_3) \rightleftharpoons CH_3F(3\nu_3) + 32 \text{ cm}^{-1} \qquad (2)$$

At $3\nu_3$ an intermode cross over step to ν_1, ν_4 is possible

$$CH_3F(3\nu_3) + CH_3F \rightleftharpoons CH_3F(\nu_1, \nu_4) + CH_3F + \sim 140 \text{ cm}^{-1} \quad (3)$$

However, the observed absolute population in the C-H stretches ν_1, ν_4 is much smaller than that in the $3\nu_3$ level. An explanation may be found in recent spectroscopic data on the region near the C-H stretch frequency in CH_3F[6]. Coriolis interactions do mix the $3\nu_3$ and C-H stretch frequencies, but only a relatively high J, K states, where the fractional population is very small (see Fig. 4). Due to this specific kind of coupling via local resonances the rise time of the 3.3 μm fluorescence may be short but the energy transfer process inefficient. The latter is in agreement with results reported in[7] where the energy stored in the individual vibrational modes of Methylfuoride following intense laser excitation has been measured.

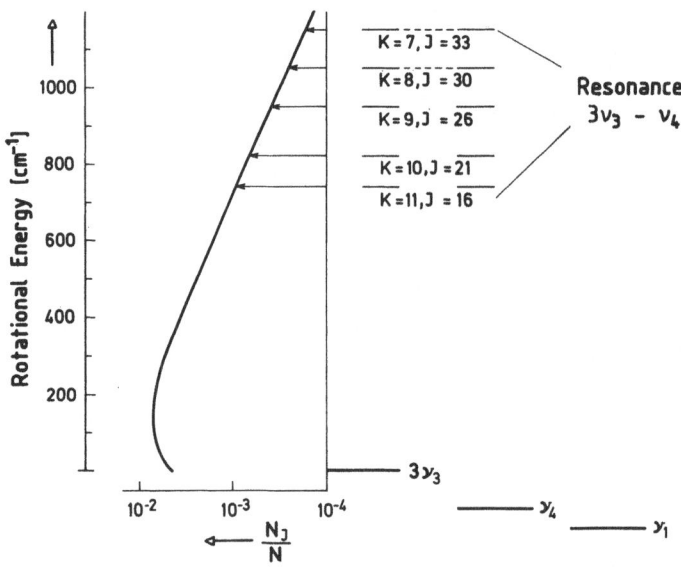

Fig. 4.   Coriolis resonances in CH_3F

In another set of experiments the time resolved fluorescence signals were studied as a function of added inert gas pressure with the CH_3F pressure kept constant at 50 mtorr. As shown in Fig. 5 Argon addition greatly increases the absolute population in the ν_1, ν_4 levels while the $2\nu_3$-population increases due to Ar addition is very small, so that vibrational energy transfer within the ν_3 manifold cannot account for the increase of one order of magnitude in the C-H stretch population. In the case of added rare gas atoms the pathways involving transfer of translational into vibrational energy

$$CH_3F(\nu_3) + Ar \rightleftharpoons CH_3F(\nu_6) + Ar - 133 \ cm^{-1} \tag{4}$$

$$CH_3F(\nu_6) + Ar \rightleftharpoons CH_3F(\nu_2, \nu_5) + Ar - \sim 280 \ cm^{-1} \tag{5}$$

become more important as can be seen by the intensity increase of the ν_2, ν_5-fluorescence. The V-V transfer process

$$2CH_3F(\nu_2, \nu_5) \rightleftharpoons CH_3F(2\nu_2, 2\nu_5) + CH_3F + 84 \ cm^{-1} \tag{6}$$

will be very efficient in populating the ν_1, ν_4 levels due to Fermi resonance between $2\nu_5$ and ν_1. Thus, in the case of Argon addition to Methylfluoride, the observed fluorescence signal of the C-H stretch vibration ν_1,ν_4 is the result of the superposition of the two different filling mechanism (1) - (3) and (4) - (6). The latter is the dominant pathway at higher Ar pressures.

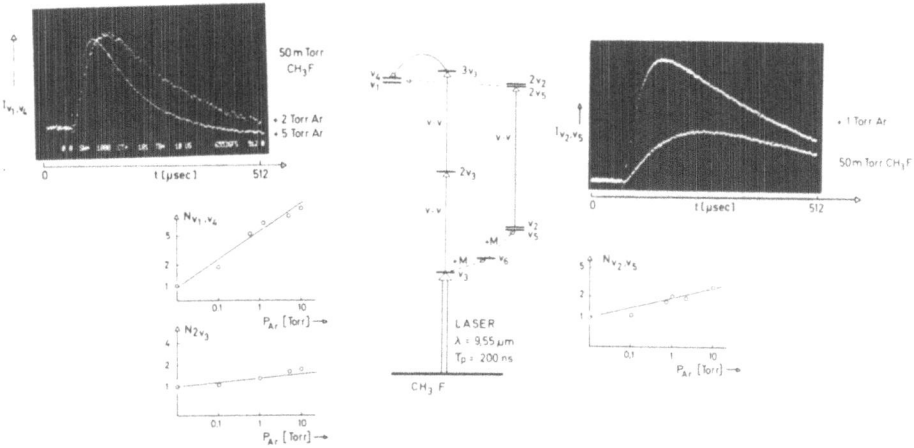

Fig. 5.   Effect of Argon addition on the population of the ν_1, ν_4 and $2\nu_3$ levels

CONCLUSIONS

The experiments on the laser induced infrared fluorescence in CH_3F molecules show that a "metastable" vibrational energy distribution with dominant population in the C-F stretching mode can be created after laser excitation. While the total vibrational energy in the excited CH_3F molecule remains nearly constant the distribution can be changed in a mode selective way by collisions with inert gas atoms. In future experiments the reaction $Br + CH_3F \longrightarrow HBr + CH_2F$ will be used as model system for a study of the mode specific energy consumption in reactions of vibrationally excited polyatomic molecules.

ACKNOWLEDGEMENT

The financial support of the Deutsche Forschungsgemeinschaft (Sonderforschungsbereich 93 "Laser Photochemistry") and the Collaboration Programm between the Hebrew University Jerusalem and the University of Göttingen is gratefully acknowledged.

REFERENCES

1. K. V. Reddy and M. J. Berry, Chem. Phys. Lett. 66, 223 (1979)
2. B. D. Cannon and F. F. Crim, J. Chem. Phys. 75, 1752 (1981)
3. H. Hippler, J. Troe and H. J. Wendelken, J. Chem. Phys., in press
4. R. S. Sheorey and G. W. Flynn, J. Chem. Phys. 72, 1175 (1980)
5. B. L. Earl, P. C. Ysolani and A. M. Ronn, Chem. Phys. Lett. 39, 95 (1976)
6. G. Graner, J. Phys. Chem. 83, 1491 (1979)
7. B. L. Earl, L. A. Gamss and A. M. Ronn, Accounts of Chemical Research 11, 183 (1978)

INFRA-RED LASER INDUCED ENERGY DISTRIBUTIONS IN

POLYATOMIC MOLECULES

R.G. Harrison

Department of Physics
Heriot-Watt University
Riccarton, Edinburgh

INTRODUCTION

The nature of the energy distribution within polyatomic molecules at low pressures subjected to multiple photon excitation by
infrared irradiation has been the subject of much controversy over
the last few years[1]. Although it is now generally accepted
that energy is not retained in the vibrational mode initially
excited it is not clear by what extent energy is randomised among
the other vibrational degrees of freedom of the molecule. Consequently the role of i.r. laser induced multiple photon dissociation (MPD) in offering any advantages over conventional pyrolitic techniques through control of reaction processes by
modification of the energy distribution remain unclear.

The unimolecular MPD of polyatomic molecules has to date, been
extensively investigated[1] in molecules which exhibit a single
dominant reaction channel. There are however considerable
advantages in studying the MPD of molecules which possess two or
more competing reactions channels, giving rise to chemically
distinct products, since the ratio of products from the individual
channels (the product branching ratio) serves as a direct
measure of the energy distribution within the excited molecule.
This approach has been adopted in the cases of ethylvinylether[2,3]
CF_2XY[4] (X,Y = Cl,Br), CH_2DCH_2Cl[5], C_2H_4FX[6,7], x = Cl[6],
Br[7], C_2H_2FCl[8], cyclobutylchloride[9], vinylcyclopropane[10]
and cyclobutanone [11a-f].

Essential to an unambiguous interpretation of the data is a
concomitant characterisation of the absorption process by which
the laser induced elevated gas temperature, over the range of

141

operating conditions employed in the reaction studies, may be determined. Specifically, precise determinations of the average number of quanta absorbed per molecule, <n>, is an important prerequisite. Unfortunately such measurements have been sparse or omitted in most of the previous investigations (cited above) for multi-channel molecules.

As part of an ongoing application of this approach to a series of molecules we summarize below results for two of these, namely cyclobutanone and 1-bromo-2-fluoroethane.

MOLECULAR SPECIES

a) Cyclobutanone

The MPD either direct or sensitized of cyclobutanone is well suited to study since accurate Arrhenius parameters are known for both molecular decomposition routes.

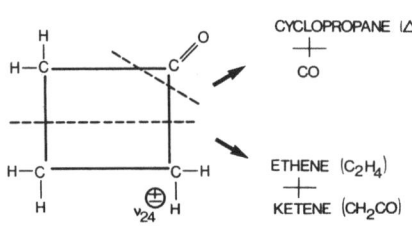

The first (E_A = 52 kcal/mode) yields ethylene plus ketene and the second (E_A = 58 kcal/mode) cyclopropane and CO plus minor amounts of propylene. Analysis of the C_2H_4 and C_3H_6 yields a direct measure of the individual channels. As no significant wavelength effects were observed from 1067 – 1086 cm^{-1}, the R(12) line (1073 cm^{-1}) was chosen to give optimum absorption by the ν_{24} fundamental mode.

b) 1-bromo-2-fluoroethane

This molecule is of particular interest since two independent vibrational modes are accessible to excitation, assigned as the C–C stretch mode at 1073 cm^{-1} (R12 9μm band of CO_2) and the CF stretch mode at 1027 cm^{-1} (P40 9μm band of CO_2). Our recent pyrolitic investigations show decomposition to occur via two well defined first order pathways with activation energies of 54 kcal/mol and 60 kcal/mol for HBr and HF elimination respectively.

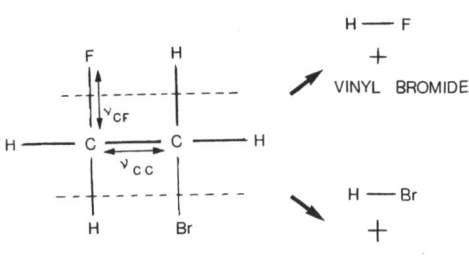

For MPD, excitation of the CF stretch mode results in a more direct energization of a bond associated with HF elimination and as such a localized energy distribution may occur.   This contrasts with that expected from excitation of the C-C stretch.

EXPERIMENTAL

A Lumonics grating tunable TEA CO_2 pulsed laser operating at 0.5 Hz with an output of up to 10 J/pulse (single line) was used as the irradiation source.   The reaction was carried out in 8cm Pyrex cells of internal diameter 25cm fitted with polished NaCl end windows.   Energy absorption was measured both conventionally by using joulmeters (Jentek ED200) and for higher sensitivity by opto-acoustic techniques.   A collimated beam geometry of beam diameter 9mm was used throughout for both reaction and absorption measurements.   Laser pulse attenuation was achieved by polythene sheets.   Reactant pressures were measured by a capacitance manometer (MKS Beratron 220).   The molecular species were used after degassing when no impurities were evident by gas chromato-graphic analyses which was also used to analyse the products formed by laser irradiation.

ENERGY ABSORPTION

For both molecules plots of optical density $\log_{10}(F_o/F_t)$, where F_o and F_t are the incident and transmitted fluence gave excellent linear dependence on gas pressure at constant F_o for a range of F_o, demonstrating absorption features consistent with Beer-Lambert behaviour under the conditions of the experiment. Furthermore the optical density was also found to be independent

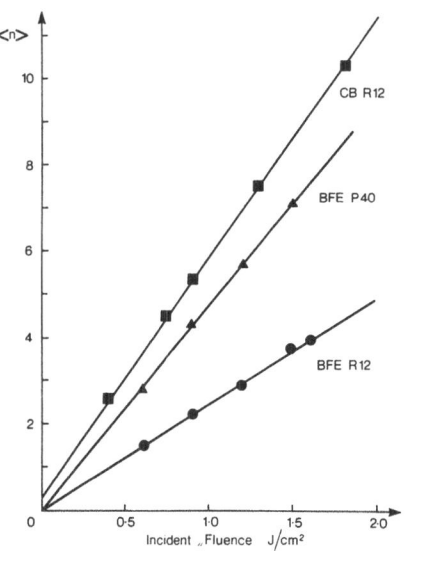

Fig.1   Plots of <n> as a function of laser fluence for fixed gas pressure for cyclobutanone (CB) and 1-bromo-2-fluoroethane (BFE).

of laser fluence F_o for constant pressure.   The absorption cross
section for the multiple photon process was thus independent of
both gas pressure and input laser fluence over the range investigated ($\sim$ 0.2 - 1.6 J/cm^2 and $\sim$ 0.5 - 10 torr for 1-bromo-2-
fluoroethene and $\sim$ 0.2 - 2.0 J/cm^2 and 0.5 - 10 torr for cyclobutanone).   Corresponding plots of <n> as a function of laser
fluence are presented in fig. 1.   Good linear dependences are
shown confirming independence of absorption cross section on
incident laser fluence for both molecules.

PRODUCT ANALYSIS

     For both molecules the fraction of reactant decomposed per
pulse in the irradiated volume was a few percent at the higher
pressures ($\geqslant$ 10 torr).   Measurements of yield per pulse were
found to be independent of the total number of pulses confirming
that all major products measured arose from primary decomposition
routes.

a) <u>Cyclobutanone</u>

     The major product analysed was ethene with lesser amounts
of cyclopropane and propene, the ratio of ethene to cyclopropane
and propene never falling below $\sim$ 30.

     The product yield ratio (R) of ethene:cyclopropane and
propene monitored as a function of pressure for three of the
fluences investigated are shown in fig. 2, the trends of which
are representative of those obtained for other fluence values.

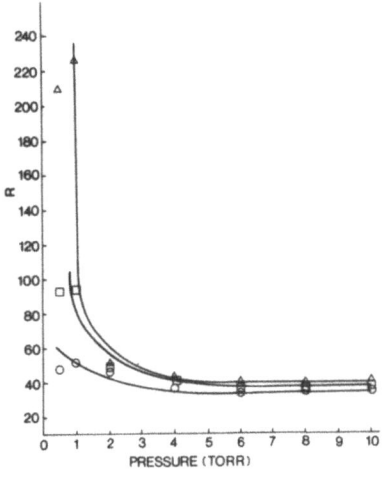

Fig. 2.   Product yield ratio R
          as a function of
          pressure of cyclobutan-
          one for fixed incident
          laser fluences F_i/Jcm^{-2}
          0, 1.98; $\square$, 1.17;
          $\triangle$, 0.90.   Theoretical
          predictions are given
          by solid lines.

Down to a pressure of ⋏ 3 torr the ratio, for a particular
incident fluence stays constant around 35, whereas at lower
pressures it rises rapidly giving highest values for lowest
incident fluences.

b) 1-bromo-2-fluoroethene

     The major products analysed were vinyl flouride with a
lesser amount of vinyl bromide in the ratio of 10:1. Minor
products from the radical chain mechanism of Br elimination were
added to the vinyl flouride yield to constituent total product
yield from the lower activation energy (HBr elimination). Plots
of R (vinyl flouride plus minor products to vinyl bromide) versus
pressure for the same absorbed energy (<n> ⋏ 7) are shown in
fig. 3 for excitation of the CF stretch and C-C stretch modes
respectively. The trends for both are essentially the same
exhibiting a decrease in product ratio by a factor of two at the
lower pressures. These results are consistent with the trends of
pyrolysis data which are attributed to pressure fall off effects.

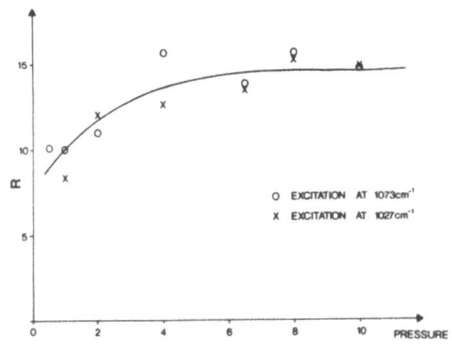

Fig. 3. Product yield ratio R
as a function of
pressure of 1-bromo-
2-fluoroethane for
excitation at 1073cm^{-1}
(O) and 1027cm^{-1}(X)
with the same absorbed
energy.

DISCUSSION

     The results for both molecules show that for any fixed laser
fluence the energy absorbed per molecule <n> (and consequently
the temperature of the gas) remained constant over the complete
pressure range.

     For cyclobutanone the constant value of product yield ratio
R down to pressures of c.a. 3 torr is consistent with a thermally
equiliberated system as confirmed in our previous work.[11a,c].
However in striking contrast to this data, at low pressures
(< 3 torr) the R rises dramatically[11c]. Thus although the average

excitation energy <n>remains constant it is preferentially directed into the lower decomposition channel. Consequently the reaction process at these pressures, (which cannot be explained by
thermal pressure fall off effects), is indicative of a molecular
energy distribution narrower than Boltzman. Particularly notable
is the pronouncement of the effect for reduced laser fluence, the
lower channel of decomposition being increasingly favoured. For
1-bromo-2-fluorethene, the product yield ratio, R. together with
its dependence on gas pressure and laser fluence for both modes
of excitation appear to be essentially the same, fig. 3 and
similar to those obtained by conventional pyrolysis.

In view of the non-thermal features exhibited in cyclobutanone even though the mode of excitation is not obviously
identified with either reaction pathway the results for 1-bromo-
2- fluoroethene are somewhat surprising. Contrary to expectat-
ions excitation of the CF stretch mode resulting in more direct
energization of the bond associated with HF elimination, does not
result in energy localization and enhancement of this channel.
The significantly different trends exhibited by these two molecules can not be obviously explained by density of state
arguments since calculations according to the method of Whitten
and Rabinovitz show the density of states at levels of excitation
near the dissociation limit are, for 1-bromo-2-fluoroethene, some
two orders of magnitude smaller than those for cyclobutanone.
Since it is considered that redistribution of population between
states occurs most probably for states of equal energy, within
the uncertainty principle, then intra molecular energy transfer
under collisionless conditions is efficient only when the density
of states is considerably greater than 10^9 cm^{-1} corresponding to
an uncertainty width equal to the reciprocal of the radiative
lifetime of a few milliseconds. Although contrary to our
findings, it could therefore be expected that localization of
energy and corresponding non-thermal reaction processes be more
probable in 1-bromo-2-fluoroethene, particularly for excitation
of the CF stretch mode, than for cyclobutanone; the density of
states at for example the lower activation energies being
$\sim 5 \times 10^9$ and 10^{11} cm^{-1} respectively for the molecules.

Although at the present time we have no explanation for these
differences an important indication of the nature of the
excitation process for these molecules is evident from the results
of energy absorbed (fig. 1). That the number of photons absorbed per molecule <n> increases linearly with laser fluence and
is independent of pressure is consistent with the multiple
photon excitation behaviour of a simple harmonic oscillator[12].
As such it may be inferred that the initially excited modes of
these molecules, although undoubtedly coupled to other vibrational
degrees of freedom by anharmonic forces, retains much of their
oscillator strength. The resultant admixture of states

nevertheless sufficiently compensates for anharmonicity of the resonantly pumped mode to give rise to multiple photon excitation processes similar to that for a simple harmonic oscillation. Interpretation of data is based on the statistics of modified SHO theory in which earlier treatments[13] are considerably extended[14,15] to account for ladder truncation and bimolecular collisional energy transfer processes, both particularly significant for gas pressures >1-2 torr. If f_v denotes the normalised excitation probability of the vibrational mode then the mean vibration number $\bar{v}$ is given by $\bar{v} = \Sigma v f_v$. The rate of change $d\bar{v}/dt$ is independent of collisions (since binary v-v processes conserve $\bar{v}$) but in terms of the radiative rate α (proportional to radiation intensity) is given by $d\bar{v}/dt = \alpha$.[14] By integration, $\bar{v} - \bar{v}(o)$, equal to the number of resonantly absorbed photons $<n>$, is αt, which in suitable units, is the fluence F. Such linear dependence, namely $\bar{v} = \bar{v}(o) + F$, is shown in the experimental data of fig. 1 as $<n> = F$.

It is assumed that dissociation 'a' occurs for $v \geqslant a$ at a rate αf_a. It follows that the product ratio is proportional to $R = (f_b - f_a)f_a^{-1}$. Since the relative change in activation energy is small, then $R = (f^{-1} df/dv)$ can be estimated approximately from the mean width of the vibrational energy distribution. Thus if $\sigma = <(v-\bar{v})^2>$, then $R \simeq \sigma^{-\frac{1}{2}}$. Considering the binary collision rate equation it can be shown that for an untruncated vibrational ladder, $\sigma - \bar{v}(\bar{v} + 1)$ decays according to the equation

$$\frac{d}{dt} [\sigma - \bar{v}(\bar{v} + 1)] + 2\gamma[\sigma - \bar{v}(\bar{v} + 1)] = o \tag{1}$$

where γ is the collisional rate constant, proportional to pressure p. The effect of ladder truncation (due to anharmonicity and dissociation) is to modify this equation to the form

$$\frac{d}{dt} [\sigma - \bar{v}(\bar{v} + 1)] + 2\gamma [\sigma - \bar{v}(\bar{v} + 1)] = -2S\tilde{\alpha}\bar{v} \tag{2}$$

where $S = 2(1 + \frac{Q'}{Q})$ represents the renormalisation, $\overset{N}{\Sigma}f_v = Q$, $\overset{N}{\Sigma}vf_v = Q'\bar{v}$, due to truncation at $v = N$ and $\tilde{\alpha} = Q\alpha$ is an effective rate. A complete solution of this equation in the high $v \gg 1$ limit is given by

$$\sigma - v^2 = \frac{\alpha S}{\gamma} (v(o) - \frac{\alpha}{2\gamma}) |\exp(-2\gamma t)-1| - \frac{\alpha^2 St}{\gamma} \tag{3}$$

For cyclobutanone in the high pressure limit ($\sim$ 10 torr), equilibriation is rapid and leads to a thermal distribution, namely $\sigma = v(v + 1) \simeq v^2$, to give $R_\infty^{-1} = 10^{-2}(2.2 + 0.36 F)$ where F is the fluence, proportional to αt (t is maintained constant and is taken as the pulse duration). Since F in these units is $\sim$ 1.0, the rise in effective vibrational temperature is about $0.36/2.2 \simeq 16\%$.

For P << 10 torr, equilibriation is incomplete and is expressed by a narrowing of the distribution. Assuming $\gamma t \gg 1$, then formula (3) gives

$$R^{-2} \doteq R_\infty^{-2} - KF \, P^{-1} \qquad\qquad (4)$$

where K includes the unknown truncation factor S, and P is kinetically related to γ by the ratio $\gamma/P = a/p$, a = collision scattering cross-section, P = mean molecular momentum. The coefficient K is determined by plotting $R^{-2} - R_\infty^{-2}$ against P^{-1}. The corresponding theoretical curves from eqn (4) of the product ratio is a function of pressure (fig. 2) are in good agreement with experiment. In the limit $P \to 0$, below the limit of our experimental measurements, eqn (3) shows that $\sigma \cdot v^2 = 2Sv(0)\alpha t$; the product ratio therefore being independent of pressure in the purely collisionless regime as expected.

For 1-bromo-2-fluoroethane the product ratio exhibited over the complete pressure range is consistant with a thermalised system. These results are interpreted by equation (1) or equation (3) with S set to unity, the case for a thermalised distribution. As such the effect of ladder truncation, the cause for non-thermal behaviour, appears to be minimal for this molecule.

CONCLUSIONS

The comprehensive study of the infrared MPD of polyatomic molecules exhibiting multiple dissociative pathways is important in elucidating their vibrational energy distribution. We have shown that measurements of both reaction yields and <n> are essential to such a study. In particular, measurements over a sufficient pressure range to encompass the transition from non-collisional to the collisional regime are important. In the absence of <n> no unequivocal statement can be made regarding the cause of the departure from a thermal distribution as indicated by relative yield data alone.

For cyclobutanone, the width of the energy distribution in the excited molecules narrows markedly at low pressures compared to the high pressure Boltzmann distribution. At any degree of excitation, the consequence is that the lower activation energy channel, namely the production of C_2H_4 and ketene, is relatively enhanced at low pressure. Furthermore this effect is most pronounced at low fluences. For 1-bromo-2-fluoroethane the energy distribution shows no pronounced departure from the thermal distribution at the high pressures for excitation of either the C-C stretch mode or the CF stretch mode. The results for both molecules are in accord with the proposed kinetic model.

The vibrational energy distribution is uniquely assessed by the measured values of $\sigma - \bar{v}(\bar{v} + 1)$ which vanishes for quasi-thermal states. For non-Boltzmann states, however, it is finite and characterised by the radiative and collisional rates, an anharmonic term as well as laser pulse duration.

ACKNOWLEDGEMENTS

Special thanks are given to P. John, M. Humphries and P.G. Harper for their full collaboration in this work.

REFERENCES

1. (a) R.V. Ambartzumian and V.S. Letokhov, Chemical and Bio-chemical Applications of Lasers, edited by C.B. Moore (Academic Press, N.Y., 1977). Vol. III p.167; (b) P.A. Schulz, Aa. S. Sudbo, D.J. Krajnovich, H.S. Kwok, Y.R. Shen and Y.T. Lee, Ann. Rev. Phys. Chem., 30, 379 (1979); (c) Multiple Photon Excitation and Dissociation of Polyatomic Molecules, edited by C.D. Cantrell (Springer, N.Y. 1979); (d) M.N.R. Ashfold and G. Hancock, Chem. Soc. Sp. Per. Rep., 4, 213 (1980).
2. D.M. Brenner, Chem. Phys. Lett., 57, 357 (1978).
3. R.N. Rosenfeld, J.I. Brauman, J.R. Barker and D.M. Golden, J. Amer. Chem. Soc., 99, 8063 (1977).
4. R.J.S. Morrison, R.F. Loring, R.L. Farley and E.R. Grant. J. Chem. Phys. 75, 148 (1981).
5. A.J. Colussi, S.W. Benson, R.J. Hwang and J.J. Tiee, Chem.Phys. Lett., 52, 349 (1977).
6. A.V. Baklanov, Yu. N. Molin and A.K. Petrov., Chem.Phys.Lett., 68 329 (1979).
7. T.H. Richardson and D.W. Setser, J. Phys. Chem. 81, 2301 (1977).
8. W.A. Jalenack and N.S. Nogar, J. Phys. Chem. 84, 2993 (1980).
9. J.S. Francisco and J.I. Steinfeld, Int.J.Chem.Kinetics, 13, 615 (1981).
10. W.E. Farneth, M.W. Thomsen, M.A. Berg, J.Amer.Chem.Soc. 101, 6468, (1979).
11. (a) C. Steel, V. Starov, R. Leo, P. John and R.G. Harrison, Chem.Phys.Lett.62,121(1979); (b) M.H. Back and R.A. Back, Can. J.Chem.,57,1511 (1979); (c) R.G. Harrison, H.L. Hawkins, R.M. Leo and P. John, Chem.Phys.Lett.,70,555(1980); (d) S.Koda, Y. Ohnuma, T. Ohkawa and S. Tsuchiya, Bull.Chem.Soc. Jap. 53, 3447 (1980); (e) V. Starov,N. Selamoglu and C. Steel, J.Phs. Chem. 85,320(1981); (f) R.A. Back, Can.J.Chem. 60,2542(1982).
12. J. Stone and M.F. Goodman, J.Chem.Phys. 71, 408 (1979).
13. J.G. Black, P. Kolodner, M.J. Shultz, E. Yablonovitch, N. Bloembergen, Phys. Rev. A, 19, 704 (1979).
14. P.G. Harper, Appl. Phys. B (submitted).
15. P. John, M. Humphries, R.G. Harrison and P.G. Harper, J.Chem.Phys. (submitted).

DYNAMICS OF MULTIPHOTON EXCITATION OF POLYATOMIC MOLECULES BY

MEANS OF ONE OR TWO IR LASER FREQUENCIES*

R. Fantoni, E. Borsella and A. Giardini-Guidoni
ENEA, Dip. TIB, Divisione Fisica Applicata - C.R.E.
C.P. 65 - 00044 Frascati, Rome, Italy

ABSTRACT

IR laser induced multiphoton excitation of polyatomic molecules (CF_3Br, C_2F_5Cl) has been investigated both in cell at room
temperature and after cooling the gas in a supersonic beam. Multiphoton resonances have been found to occur in the coherent pumping
of the first discrete levels both in single and two frequency spectra. Spectral structures have been observed in the quasi-continuum
of multiple-photon excited molecules.

I INTRODUCTION

In recent years the process of multiple-photon excitation
(MPE), and dissociation (MPD) of polyatomic molecules by IR laser
radiation has been the subject of much discussion in the scientific community [1]. It is customary [2] to divide the process of
MPE into three steps. The first stage is the coherent pumping of
the discrete non equidistant energy levels of the IR active vibrational mode. Its most characteristic feature is the occurrence of
resonances whenever the laser frequency is equal to 1/N of the energy difference between the ground state and the level with N
quanta. As the vibrational energy increases, the oscillator
strength of the pumped mode is spread over a quasi-continuum of
states and resonant absorption of laser photons is possible up to
the region of the true continuum of states above the first dissociation threshold. Although some light has already been thrown on
the whole excitation process, it is still the subject of controversy whether the quasi-continuum of polyatomic molecules is a
thermal bath [3] or a clustering of states near the harmonics of
the pumped mode [4]. Relevant role in the formation of these ag-

gregates should be played by the overtones and combination bands
almost degenerate with the levels of the excited mode. The occurrence in the quasi-continuum of structures related to the linear
absorption spectrum could support the second hypothesis. Aim of
our work is to investigate the most relevant aspects of the multiphoton excitation process, such as resonances appearing in climbing up the discrete level ladder, and to obtain information about
the "quasi-continuum" (q.c.) of states, which has to be transvers-
ed by the molecules before reaching the dissociation threshold.
The presence of narrow resonances in the multiple-photon excitation
of spherical top molecules (SF_6) has already been shown [5] in
spectra measured at low temperature. Their assignment [6] was accomplished taking into account the anharmonic splitting [7] of vibrational levels in the octahedral force field. The vibrational
levels of less symmetric molecules do not present such a splitting
and different effects must compensate their anharmonicity in order
to permit climbing up the vibrational ladder. Important contributions from rotational sublevels and vibrational hot bands are expected at room temperature. Results obtained with a line tunable
CO_2 laser on C_2F_5Cl [8] and CF_3Br [9] have shown the presence of
unresolved structures roughly corresponding to the vibrational
band heads of one- two- and three- photon resonances. This fact
led us to measure in more detail the MPE spectra of these polyatomic molecules. Experimental results are reported and discussed in
Sect. II. Investigation of the quasi-continuum has been performed
by using the two frequency multiphoton absorption (MPA) and MPD
technique, in which the first laser (ω_1) is tuned at a frequency
almost resonant with a IR active molecular vibration, while the
second laser emission (ω_2) is varied over the whole spectrum. Only
weak structures probably related to features of the quasi-continuum in the two frequency MPA and MPD experiments performed on SF_6
[10] and OsO_4 [11] were observed. In these cases the clustering of
levels in the q.c. has been ascribed mainly to anharmonic splitting of highly degenerate vibrational states. The same phenomenon
cannot play any role in less symmetrical molecules, such C_2F_5Cl
and CF_3Br. Data are shown and discussed in Sect. III.

II SINGLE FREQUENCY MPA EXPERIMENTS

a) Experimental

The infrared radiation was provided by a pulsed multimode
continuously tunable CO_2 laser (Lumonics mod. TE 281). Continuously tunable emission with a resolution of 0.12 ± 0.02 cm^{-1}, a pulse
energy up to 500 mJ and a pulse width of $\cong$ 90 nsec (FWHM) and a
tail of 400 nsec were obtained by operating the laser at 9 atm.
Multiphoton absorption was measured by using a previously described optoacoustic cell [12]. The calibration of the optoacoustic
signal was accomplished by means of conventional measurements of
the incident and transmitted laser intensities. The measurements

were performed at a pressure of 0.5 torr in the cell at room temperature. A TMS 9900 based microcomputer controls [13] the laser
frequency and fluence as well as gathers valid data (e.g. data corresponding to ± 10% of the desidered fluence).

b) C_2F_5Cl

The C_2F_5Cl molecule is a C_s asymmetric rotor having 18 nondegenerate normal vibrational modes. The v_4 normal mode is centered
near 983.0 cm^{-1}. MPA measurements performed with a line tunable
CO_2 laser (Fig. 1 inset) show the presence of broad spectral features whose structures are varying with the fluence. Many more
peaks appear when the measurements are performed with the continuously tunable CO_2 laser at a fluence of 0.3 J/cm^2 and T = 300 °K
(Fig. 1). The structures peaking around 975 cm^{-1}, whose behaviour
as a function of fluence is found to be non linear (of the order
of I^2), may be due to two or three photon resonances. Simple mode
calculation indeed show contributions of two and three photon resonances in this spectral range, superimposed on a background of
incoherent absorption. The characterization of the peaks appearing
in the spectrum will be completed when high-resolution linear absorption spectra allowing for the determination of all the molecular parameters [14] will be fully analyzed.

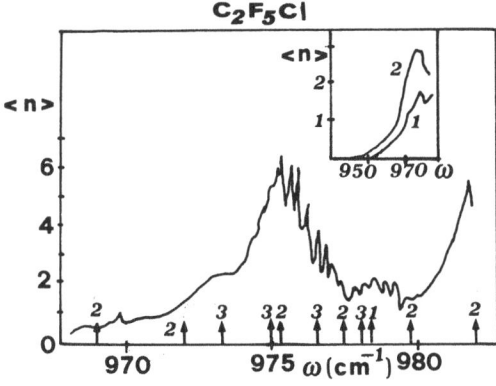

Fig. 1 MPA spectrum of C_2F_5Cl measured at ϕ=0.3 J/cm^2, p=0.5 torr and T~300 K. Arrows mark the one- two-, and three-photon resonances. In the inset MPE spectra measured with a line tunable la-ser at p = 0.2 torr, ϕ=0.03 J/cm^2 (1) and ϕ=0.2 J/cm^2 (2) is shown

c) CF_3Br

The CF_3Br molecule is a distorted tetrahedron of C_{3v} symmetry,
having six nondegenerate normal vibrational modes [15]. The
$^{12}CF_3Br$ v_1 normal mode, is a parallel vibration centered near
1084.6 cm^{-1} [15]. ^{79}Br and ^{81}Br are present at natural abundance
($\cong$ 50.5% and 49.5% respectively) giving rise to an isotope shift
of 0.24 cm^{-1} [16]. MPA spectra of CF_3Br measured with a line-tunable CO_2 laser as a function of the laser wavelength (Fig. 2
inset) show several peaks which can be interpreted as multiphoton
resonances. In particular it has been found that the heat-bath-
-feedback model proposed by Stone and Goodman [17] can account rea

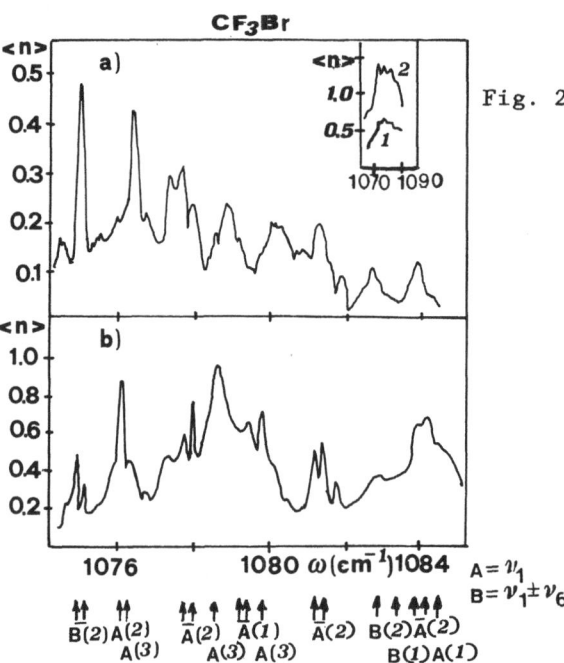

Fig. 2 MPA spectrum of CF_3Br. (a) Measurements with a continuously tunable laser at $\phi = .35$ J/cm^2, p=0.5 torr and T~300 K. (b) Calculation (see text). Arrows mark the one-, two-, and three-photon resonances. In the inset MPA spectra measured with a line-tunable laser at p=0.5 torr $\phi = 0.5$ J/cm^2 (1) and $\phi=0.8$ J/cm^2 (2) is shown

$A = \nu_1$
$B = \nu_1 \pm \nu_6$

sonably for data [18]. In this model a coherent ladder of four or five discrete vibrational levels in the ν_1 mode, followed by an incoherent absorption in the quasi-continuum is considered. Measurements performed with the continuously tunable CO_2 laser (Fig.2a) show a well resolved structure in which most of the peaks observed can be assigned as one-, two- and three- photon resonances in the ν_1 vibrorotational ladder. However the structure observed at high resolution is richer than the calculated one, and to account for all the experimental peaks it is necessary to include the absorption of the hot bands $\nu_1 + \nu_6 - \nu_6$, $\nu_1 + 2\nu_6 - 2\nu_6$, $\nu_1 + \nu_3 - \nu_3$ which are appreciably populated at room temperature [16,19]. The spectrum reported in Fig. 2b has been calculated by using the Stone and Goodman model for each independent vibrational ladder originating at $\varepsilon = 0$ for the ν_1, at ν_6 for $\nu_1 \pm \nu_6$, at $2\nu_6$ for $\nu_1 \pm 2\nu_6$ and at ν_3 for $\nu_1 \pm \nu_3$ respectively. Rotational sublevels have been included according to the expression $\varepsilon_{rot} = B_{eff} J (J+1)$, in which the centrifugal distortion and the variation of B with vibrational excitation are neglected on the basis of known linear spectroscopic data [16]. Since all of those are parallel bands [19], the selection rules for parallel vibrations of a symmetric top have been used. Furthermore only rotational transitions involving the most populated rotational sublevels ($J = J_{max} = 38$ at T = 300 K) have been considered. In the resulting spectrum the weight of the population of each hot band [16] with respect to the ground state is taken into account. Arrows at the bottom of the figure indicate the relevant multiphoton resonance corresponding to each calculated peak.

III TWO FREQUENCIES MPA AND MPD EXPERIMENTS

a) <u>Experimental</u>

In this section the results of two frequency MPA and MPD experiments on C_2F_5Cl and CF_3Br will be presented. The MPA measurements were performed with the already mentioned optoacoustic cell by using as radiation sources two line tunable CO_2 lasers equipped with low jitter units (Lumonics mod. 540K) which allow for a synchronization within 50 nsec. The apparatus for MPD experiments consists of a pulsed supersonic molecular beam which after having crossed one or two synchronized line tunable CO_2 laser beams, is detected by a quadrupole mass spectrometer placed at forward angle with respect to the molecular beam itself [20]. The MPD process has been monitored as a decrease in the primary beam intensity in the mass spectrometer as a function of the fluence and frequency of the irradiating lasers [20].

b) <u>C_2F_5Cl</u>

In MPA experiments, the frequency ω_1 of the first laser coincides with the linear absorption maximum of the ν_4 mode of C_2F_5Cl while the frequency ω_2 of the second laser is varied. The intensity ϕ_1 of the first laser is high enough to bring the excited molecules up to the quasi-continuum. Data reported in Figure 3 show the occurrence of an intense structure in the two-frequency spectrum at ω_2 shifted to the red with respect to the linear maximum. this feature has been found quite sensitive to the value of the second laser fluence [21]. In order to understand the origin of this peak we have performed both single and two-frequnecy MPD measurements under collision-free conditions in the molecular beam apparatus. The striking feature of results reported in Fig. 4 is the appearance of intense structures at frequencies ω_2 for which no appreciable MPD was detectable in the single-frequency spectrum. These peaks are evidence of strong spectroscopic structure in the quasi-continuum which might arise from intensity borrowing, due to anharmonic effects by combination bands lying near

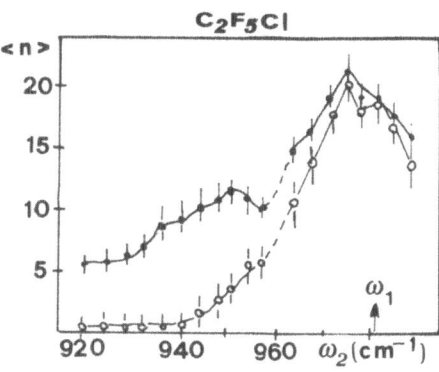

Fig. 3 MPA spectra of C_2F_5Cl measured with line tunable lasers at p=0.5 torr and T = 300 K: (o) single frequency ϕ = 1.7 J/cm^2, (•) two frequencies ϕ_1 = = 1.1 J/cm^2, ω_1=981 cm^{-1}, ϕ_2 = 1.7 J/cm^2

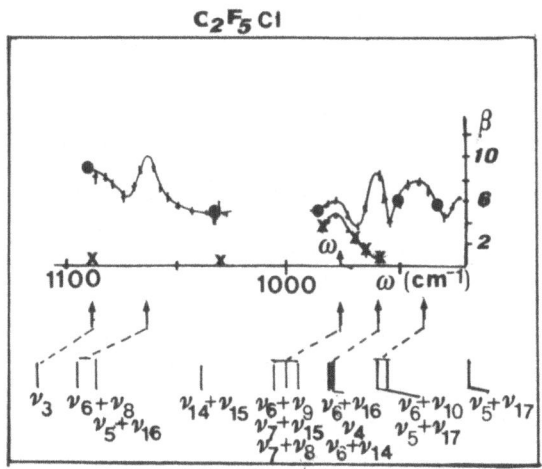

Fig. 4 MPD spectra of C_2F_5Cl measured in molecular beam: (x) single frequency $\phi=2.2$ J/cm^2, ($\bullet$) two frequencies $\phi_1 = 1.4$ J/cm^2, $\omega_1 = 977.2$ cm^{-1}, $\phi_2=2.2$ J/cm^2. Assignment of MPD structure is shown on the bottom

the I.R. active ν_4 mode [22]. A tentative assignment of MPD features to combination bands revealed by matrix-isolation spectroscopy is shown at bottom of Fig. 4. This conclusion may applay to other asymmetrical molecules and may help provide spectroscopic basis for understanding the MPE of molecules other than spherical top.

c) CF_3Br

Single and two-frequency MPA spectra of CF_3Br taken at different fluences of the first exciting laser are shown in Fig. 5. The broad structures appearing in the single MPA spectrum in the range 1067-1090 cm^{-1} are reproduced by numerical calculations based on excitation in the ν_1 vibrorotational ladder [18]. In the two color experiments the structures interpreted as multiphoton transitions starting from the first excited level are clearly enhanced, thus

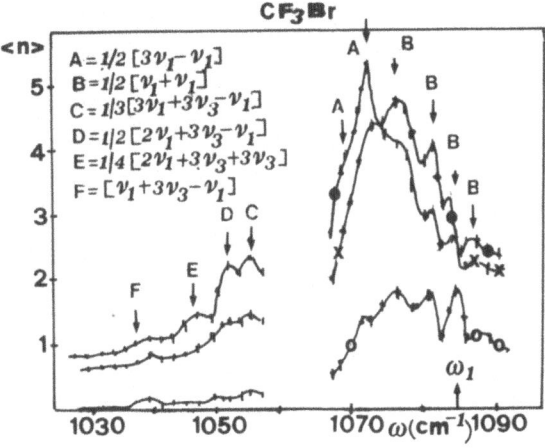

Fig. 5 MPA spectra of CF_3Br measured with line tunable lasers at p=0.5 torr and T = 300: (o) single frequency $\phi = 1.2$ J/cm^2. Two frequencies $\omega_1 = 1084.6$ cm^{-1}, $\phi_2 = 1.2$ J/cm^2: $\phi_1=0.4$ J/cm^2 (x), $\phi_1=0.8$ J/cm^2 ($\bullet$).

confirming their assignment. The peaks emerging at higher ϕ_1 should correspond to multiphoton resonances of higher order up to 1070 cm^{-1}. Mixed ladders involving the ν_1 and $3\nu_3$ overtone levels ($3\nu_3$ = 1028 cm^{-1}) could explain the occurrence of peaks in the region 1057-1027 cm^{-1}. Arrows shown in Fig. 5 mark the maximum excitation pathways involving these vibrorotational states. Also in this case, resonances starting from the first excited ν_1 leves become more prominent as the first laser fluence ϕ_1 increases. An extensive investigation of collision free MPD of CF$_3$Br has been also undertaken. Single frequency data measured at ϕ = 4.0 J/cm^2 for two different sets of temperatures [9] are shown in Fig. 6. In the spectrum at lower temperatures (Fig. 6a) two and three photon resonances in the ν_1 mode are clearly indentifiable as well as the two photon transition 1/2 ($3\nu_1$ - ν_1) starting from the first ν_1 excited state. No evidence of the one photon ν_1 resonance is found in dissociation measurements. This could be due to the anharmonic bottleneck not allowing the excited molecules to reach the continuum and to dissociate. The feature appearing at about 1086 cm^{-1} can be assigned as the two photon transition in the mixed ladder of ν_1 and $2\nu_5$ overtone in Fermi resonance between them [15, 16]. The fairly simple description of the MPE is lost when rotational and vibrational temperatures increase (Fig. 6b). Structures appear broader and not fully resolved, and also a hot band involving the ν_6 and the above mentioned mixed ladder is found. As for the fairly intense structure labelled as 1/2 ($3\nu_1$ - ν_1) in Fig. 6a

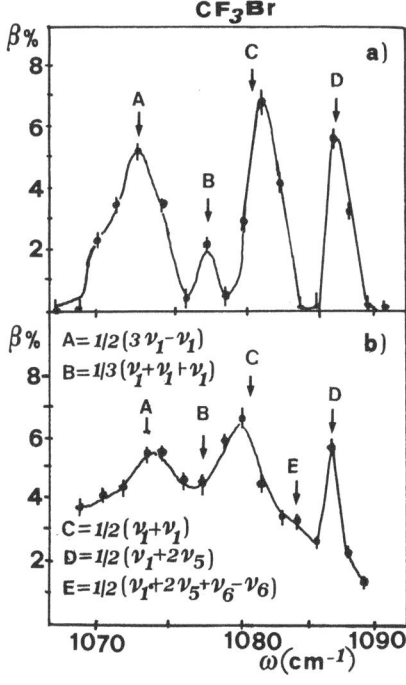

Fig. 6 Single frequency MPD spectra of CF$_3$Br measured in supersonic molecular beam at ϕ=4.0 J/cm^2. (a) 1% CF$_3$Br in Ar; (b) CF$_3$Br pure beam.

some doubt can still remain about its assignment. Being the level
v v_1 = 1 not appreciably populated when the molecule are vibrationally cold, it could be supposed that the v_1 first excited
level is populated during the laser pulse itself so originating
this hot band. In order to check this last hypothesis and to investigate the quasi-continuum structure of CF_3Br, two frequency MPD
experiments have been performed on the molecular beam. Figure 7
shows the two frequency MPD spectrum, in which the structures observed in single frequency spectra in the range 1070 - 1090 cm^{-1}
are still preserved. Furtherly the vibrational heating in the v_1
mode, supplied by the first laser results particularly effective
in increasing the intensity of the peak labelled as 1/2 ($3v_1 - v_1$),
thus confirming its assignment [23]. As far as it concerns the red
portion of the two frequency MPD spectrum (in the range 1033 -
1058 cm^{-1}), we can notice that the dissociation yield is still relevant at frequencies w_2 strongly red shifted with respect to the
linear absorption maximum.

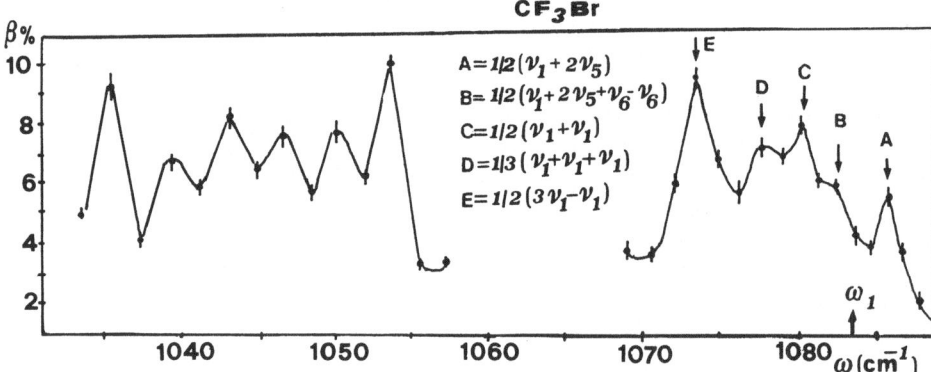

Fig. 7 Two frequency MPD spectrum of CF_3Br measured in a supersonic molecular beam of 10% CF_3Br in Ar. ϕ_1 = 1.5 J/cm^2, w_1 =
= 1084.76 cm^{-1}, ϕ_2 = 3.5 J/cm^2.

ACKNOWLEDGEMENTS

     Helpfull discussion with Prof. C.D. Cantrell, Prof. J. Reuss
and Dr. M. Dilonardo are gratefully acknowledged, as well as the
help of Dr. D. Adams, Dr. D. Masci, Dr. A. Palucci, U. Del Bello
and A. Ferretti in the experimental work. Thanks are due to R.
Belardinelli, P. Cardoni, I. Cenciarelli, M. Nardelli, S. Ribezzo
and G. Schina for technical assistance.

FOOTNOTES AND REFERENCES

*   Research supported by the ENEA (Italy), CNR (Italy) and NATO
      (Grant No. 018.82)

[1]   For a review, see C.D. Cantrell, S.M. Freund, J.L. Lyman pp.
      485-576 in The Laser Handbook, Vol. III, M.L. Stitch ed
      (Amsterdam, North Holland, 1979)
[2]   C.D. Cantrell, V.S. Letokhov and A.A. Makarov, in Coherent
      Nonlinear Optics: Recent Advances, M.S. Feld and V.S. Letohhov
      eds. (Berlin, Springer-Verlag 1980), pp. 165-269
[3]   N. Bloembergen, Opt. Commun. 15, 416 (1975)
[4]   N.V. Karlov, Invited paper, Proceedings of the 2nd Int. Conf.
      on Multiphoton Processes, Budapest (1980) eds. M. Jànossy and
      S. Varrò, p. 149
[5]   S.S. Alimpiev, N.V. Karlov, S.M. Nikiforov, A.M. Porkhorov,
      B.G. Sartakov, E.M. Khokhlov and A. Shtarkov, Opt. Commun. 31,
      309 (1979)
[6]   M. Dilonardo, M. Capitelli and C.D. Cantrell, Applied Phys.
      B29, 181 (1982); D.P. Hodgkinson, A.J. Taylor, D.W. Wright
      and A.G. Robiette, Chem. Phys. Lett. 90, 230 (1982)
[7]   C.D. Cantrell and H.W. Galbraith, Opt. Commun. 18, 513 (1976)
      and 21, 374 (1977)
[8]   E. Borsella, R. Fantoni and A. Giardini-Guidoni, Proc. of 2nd
      GNEQP Palermo 1980
[9]   E. Borsella, R. Fantoni, A. Giardini-Guidoni, D. Masci, A.
      Palucci and J. Reuss, Chem. Phys. Lett. 93, 523 (1982)
[10] E. Borsella, R. Fantoni, G. Petrocelli, G. Sanna, M. Capitelli,
      M. Dilonardo, Chem. Phys. 63, 219 (1981)
[11] R.V. Ambartzmian, V.S. Letokhov, G.N. Makarov and A.A.
      Puretzky, Opt. Commun. 25 69 (1978)
[12] G. Sanna, M. Nardi and M. Bernardini, Proc. Int. Conf. on
      Lasers 1981 (C.B. Collins, ed.) (STS Press, 1982) p. 83
[13] E. Borsella, R. Fantoni, A. Giardini-Guidoni, D.R. Adams and
      C.D. Cantrell Chem. Phys. Lett. (submitted)
[14] S. Nunziante Cesaro, M. Maltese and C.D. Cantrell to be
      published
[15] A. Baldacci, A. Passerini and S. Gherzetti, J. Mol. Spectr.
      91, 103 (1982)
[16] H. Bürger, K. Burczky, P. Shulz and A. Ruoff, Spectrochimica
      Acta 38A, 627 (1982)
[17] J.A. Horsley, J. Stone, M.F. Goodman, D.A. Dows, Chem. Phys.
      Lett. 66, 461 (1979) and references therein
[18] E. Borsella, D. Masci, M. Capitelli and M. Dilonardo, Chem.
      Phys. (1983 in press)
[19] K. Burczyk, H. Bürger, A. Ruoff and P. Pinson, J. Mol. Spectr.
      77, 109 (1979)
[20] E. Borsella, R. Fantoni, A. Giardini-Guidoni, Int. J. of Mass.
      Spectrometry and Ion Physics, 47, 81 (1983)
[21] E. Borsella, R. Fantoni, A. Giardini-Guidoni, Chem. Phys.
      Lett. 87, 313 (1981)
[22] E. Borsella, R. Fantoni, A. Giardini-Guidoni, C.D. Cantrell,
      Chem. Phys. Lett. 87, 284 (1982)
[23] A. Giardini-Guidoni, E. Borsella, R. Fantoni, Gazzetta Chimi-
      ca Italiana (1983 in press).

MULTIPHOTON IONIZATION AND FRAGMENTATION OF

POLYATOMIC MOLECULES

F. Rebentrost

Max-Planck-Institut für Quantenoptik
8046 Garching, Germany

INTRODUCTION

Multiphoton ionization (MPI) in intense visible or UV laser fields has many interesting aspects as a source of information about energetic polyatomic molecules (ions)[1-11]. An important feature is that by adjusting laser intensity and wavelength one has a rather complete control over the extent of the fragmentation accompanying MPI. Several factors make MPI of interest for mass spectroscopy. Among these are the effective ionization by lasers and the possibility of two-dimensionsional (mass and laser wavelength) mass spectrometry. Also high selectivity with respect to conventional mass spectrometry is obtained and useful e.g. to distinguish between isomers of organic compounds[12]. MPI has also interesting applications in condensed phases, e.g. ionization studies in liquids[13,14] or laser desorption of molecules from surfaces[15].

MPI of molecules in the gas phase has progressed along two lines. For rather small molecules (triatomics like NO_2, H_2S) the ionization and fragmentation of selected intermediate states have been investigated in detail[16-18]. For the case of larger polyatomic molecules like benzene (Fig. 1) some more general features in the ionization and fragmentation behavior become apparent. It is this case that shall be discussed here together with theoretical models for MPI fragmentation [19-23] which assume a statistical flow of energy.

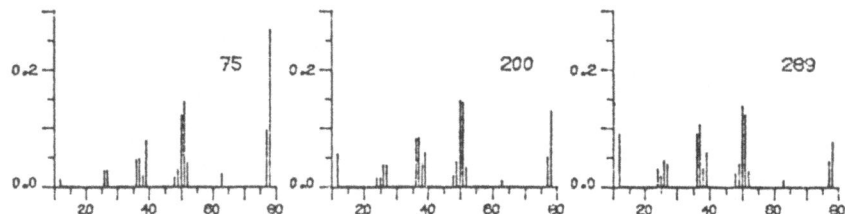

Fig. 1: MPI mass spectra for benzene by KrF laser excitation
(248 nm) at various laser intensities (in MW/cm^2)[4].

MECHANISTIC ASPECTS

With typical ionization potentials (IP) of atoms and molecules around 10 eV, at least two to four photons in the visible or
UV are required for ionization from the ground state. In the multiphoton ionization of atoms the ionization rate commonly follows
a power law of the form I^n indicating a true n-photon ionization.
On the other hand the rate of MPI of polyatomic molecules is almost always determined and resonantly enhanced by the absorption
to an excited electronic level (multistep photon ionization).
This resonant enhancement of MPI (REMPI) makes the overall ionization very efficient. This is particularly true for MPI by UV lasers when the resonant excited state is reached by one photon. A

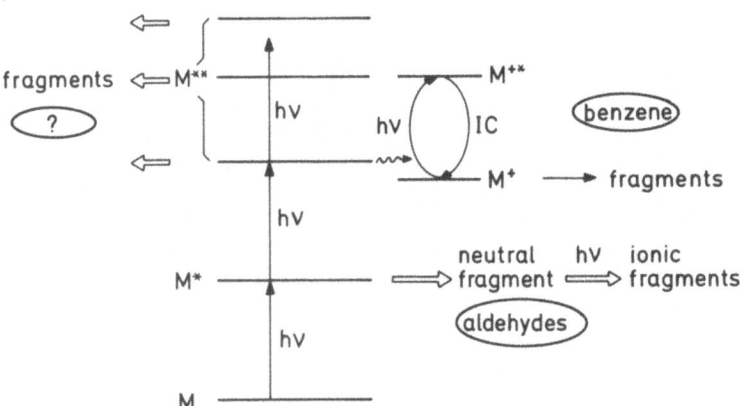

Fig. 2: Mechanistic routes for MPI fragmentation (scheme).

few typical conditions for MPI by e.g. a KrF laser are: photon energy: 5 eV; wavelength: 248 nm; pulse length: 20 ns; intensity (focussed): $\leq$ 300 MW/cm^2; absorption rate at a cross section of 10^{-17} cm^2: 20 eV/ns = 4 photons/ns.

At present three routes causing ionization are discussed frequently (see Fig. 2). These are influenced by the relative rate of the pumping by the lasers and of the internal processes in the molecule (photodissociation, radiationless transitions, autoionization).

Neutral photodissociation mechanism

Here a resonantly excited state of the neutral undergoes dissociation to smaller neutrals. Ions are formed from these fragments by consecutive absorption of photons, but no or few parent ions are observed in the mass spectra. In general one expects this mechanism, which applies to some of the aldehydes (rapid CO elimination), when the rate of dissociation exceeds the absorption rate in the excited state. Therefore increasing the rate of photodissociation by larger excess energy in the excited state (e.g. by blue tuning) or decreasing its absorption rate (by lowering the laser intensity) favor a mechanism involving neutral photodissociation steps. For example in benzaldehyde MPI at 355 nm leads to parent ions, while at 266 nm primarily $C_6H_5^+$ is observed[24]. In 248 nm excitation of acetaldehyde (IP = 10.22 eV) and butyraldehyde (IP = 9.8 eV) only the latter can be ionized by two photons. This 2-photon ionization rate seems to be fast enough to compete with dissociation and parent ions are seen in the mass spectrum. A low rate of 3-photon ionization favors dissociation from an excited state with the resulting in no parent ions in the case of acetaldehyde[3]. Of course a neutral photodissociation step need not be limited to the parent and eventually most of the laser energy is used to form neutrals rather than ions[25].

Parent ion fragmentation mechanism

Here pumping the neutral with a number n of photons sufficient to reach the IP is followed by formation of parent ions. The last photon leads either to an autoionizing state or directly into the ionic continuum. Accordingly one expects the kinetic energy of the electrons released to be less $nh\upsilon$-IP, as has been demonstrated in several cases[23,26]. The parent ion mechanism is frequently realized and most likely applies to benzene and other organic hydrocarbons. Since the ions are initially formed with little internal energy further absorption is required in the ion for effective fragmentation. Absorption in the ions may be rather favorable because of low lying electronic states (caused by unfilled valence orbitals). Also rapid conversion of electronic to nuclear energy (radiationless transitions with rates of 10^{12} s^{-1})

may be typical for larger polyatomic ions. Under these circumstances the statistical theory of unimolecar decay should be applicable to MPI.

Autoionization ladder mechanism

It is conceivable that the (autoionizing) states of the neutral continue to absorb photons leading to superexcited molecules which form the fragments. This mechanism requires that the pumprate exceeds autoionization and dissociation rates in the superexcited state (or its predecessor). Autoionization occurs typically within 10^{-13} s in which case pumping is too slow. Though initially an autoionization mechanism was promoted for MPI no example can be given at present. The case of H_2S has been thought to fall into this category for a while[17,18].

Simultaneous vs. consecutive fragmentations

On the average a molecule (ion) can absorb photons until its fragmentation rate is approximately equal to the rate of absorption of another photon. As seen from Fig. 3 the fragmentation rate (within the RRKM approximation) is rapidly increasing with excess energy. For typical pumping with lasers of ns pulse duration maximum excess energies are of the order of 10 eV. In most cases this energy is not enough to explain all the observed

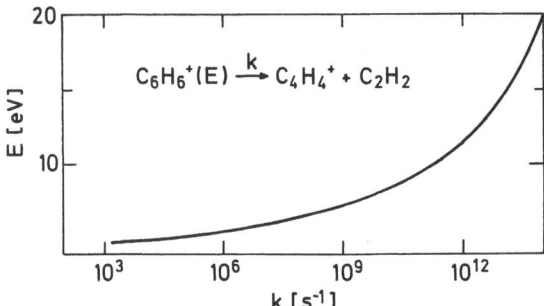

Fig. 3: Excess energy and RRKM rates for unimolecular decay of $C_6H_6^+$ ions to $C_4H_4^{+}$ [22].

fragments, with the two important consequences that fragmentation involves consecutive steps and further some of the fragments must

also absorb energy (ladder switching). In this case these intermediates must be formed within the few ns pulse duration.

Also the reported inefficiency of ultrashort pulses to bring
about extensive fragmentation may be due to some internal time
lag in dissociation or internal relaxation[28]. Further evidence
for a consecutive mechanism of fragment formation is obtained
from the kinetic energy releases of the ions[27,29].

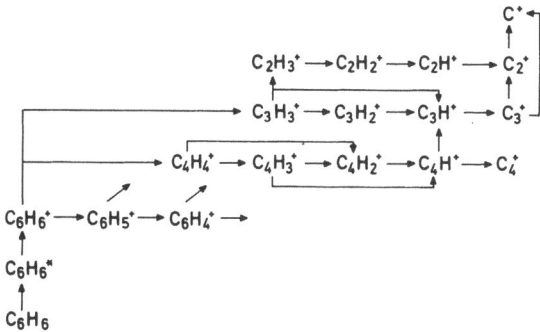

Fig. 4: Schematic reaction tree for benzene.

Fragmentation pathways

Also in MPI certain similarities of the kind and abundancies
of the ions with respect to other ionization methods (VUV photoionization, electron impact, charge transfer) are observed. Similar ions may for instance be seen in electron impact mass spectra
and MPI at medium laser intensities. This could be indicative of
similar mechanisms and the fragmentation follows along some tree
of consecutive and parallel reactions (Fig. 4). Of course MPI is
rather unique due to the possibility of very extensive fragmentation at high laser intensities. Similar to electron impact the
fragmentation of metastables can lead to direct evidence for mechanistic pathways[30].

It is of interest that often the neutrals formed by fragmentation are not ionized themselves by MPI. This has been shown by using delayed pulses of suitable wavelength (giving rise only to fragmentation without increasing the total number of ions) or sweeping out the ions from the laser focus by an applied electric field[31,32]. A reason for this may be that most of the typical neutrals (H, H_2, C_2H_2) are difficult to ionize under MPI conditions.

STATISTICAL MODELS OF MPI FRAGMENTATION

Models with mechanistic assumptions

According to the above discussion the MPI fragmentation occurs by multiple absorption/fragmentation (MAF) steps along a tree of consecutive and parallel reactions. The statistical theory of unimolecular reactions provides expressions for the fragmentation rates, branching ratios and energy distributions (Tab. 1) that can be used in a kinetic model[22,33].

Tab. 1: Quantities characterizing a fragmentation step.

$$i^+(E_i) \rightleftarrows \quad j^+(E_j) + n(E_n = E_i - E_{ij} - E_j)$$

Rates

$$k_{ij}(E_i, E_j)$$
$$k_{ij}(E_i) = \int dE_j \, k_{ij}(E_i, E_{ij})$$
$$k_i(E_i) = \sum_j k_{ij}(E_i)$$

Branching fraction

$$F_{ij}(E_i) = k_{ij}(E_i)/k_i(E_i)$$

Energy distribution in fragment j

$$f_{ij}(E_i, E_j) = k_{ij}(E_i, E_j)/k_{ij}(E_i)$$

Energy-specified probability

$$P_{ij}(E_i, E_j) = F_{ij}(E_i) f_{ij}(E_i, E_j)$$
$$= k_{ij}(E_i, E_j)/k_i(E_i)$$

Depending on the assumptions about the transition state (tight or loose) the expressions differ somewhat. For the products phase space model (based on product-like transition states) one has e.g. for the probabilities P_{ij}, in terms of the density-of-states ρ_i, ρ_n of the products

$$P_{ij}(E_i, E_j) = const \, \rho_j(E_j) \, \rho_n(E_n)$$

A kinetic description of the fragmentation is then easily conceivable taking into account also the gain in energy by absorption. The problem arises from the absorption cross section for intermediates which are unknown or do not refer to the energy-rich species involved. In two limiting cases a time-independent description can be obtained and the problem of intermediate absorption treated in a global manner. These involve a) either no absorption in any of the daughter ions[21,22] (absorption/multiple fragmentation (AMF) step model) or b) constant absorption in all ions[23].

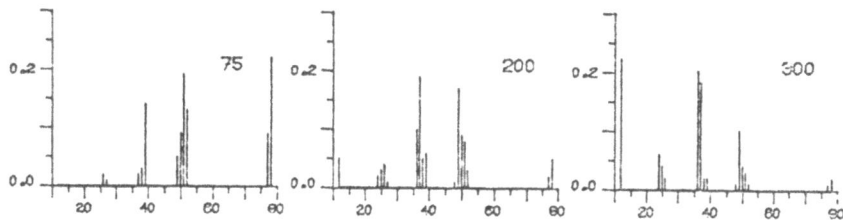

Fig.5: AMF phase space model calculations for MPI mass spectra of benzene at various laser intensities (in MW/cm^2)[22].

Maximum entropy model

The entropy function chosen (Tab. 2) is appropriate for a system of independent (fragmenting) molecules. The particle numbers X_j and quantum state probabilities x_{ij}. then result by maximizing this expression subject to a number of constraints. Evident constraints are due to the normalization of the probabilities and to the conservation of the atoms and charges constituting the species.

Tab. 2: Maximum entropy model[19,20].

Entropy:

$$S = -R\sum_j X_j \ln X_j + \sum_j X_j S_j + R\,X\,\ln X \qquad (S_j = -R\sum_i x_{ij} \ln x_{ij})$$

Constraints:

probability	$\sum_i x_{ij} = 1$	total number	$\sum_j X_j = X$
atoms + charge	$\sum_j a_{kj} X_j = C_k$	energy	$\sum_j X_j \sum_i \varepsilon_{ij} x_{ij} = \langle E \rangle$

Solution:

$$X_j = X\,Q_j(\beta)\,\exp(-\sum_k \gamma_k a_{kj}) \qquad x_{ij} = \exp(-\beta\varepsilon_{ij})/Q_j(\beta)$$

The only "physical" constraint postulated is then that the variable pumping conditions are reflected by giving a mean energy to the system. Particular mechanistic routes that have already been established in a number of cases can be implemented in principle by adding further constraints to the above "most" statistical model.

In general the models predict mass spectra in qualitative agreement with experiment (Fig. 5). Additional comparison can be made on the energy releases of the products[28,34]. One concludes that the fragmentation behavior of the larger polyatomics is primarily determined by statistical redistribution of the energy. Of course deviations from statistical behavior will almost certainly occur, as is well known in unimolecular decay.

REFERENCES

1   L. Zandee and R.B. Bernstein, J. Chem. Phys. 70, 2574 (1979); 71, 1359 (1979).
2   U. Boesl, H.J. Neusser and E.W. Schlag, J. Chem. Phys. 72, 4327 (1980).
3   J.P. Reilly and K.L. Kompa, Advances in Mass Spectr. Vol. 8 (1979).
4   J.P. Reilly and K.L. Kompa, J. Chem. Phys. 73, 5468 (1980).
5   G.J. Fisanick, T.S. Eichelberger IV, B. A. Heath and M.B. Robin, J. Chem. Phys. 72, 5571 (1980)
6   D.M. Lubman, R. Naaman and R.N. Zare, J. Chem. Phys. 72, 3034 (1980).
7   M. Seaver, J.W. Hudgens and J.J. Decorpo, Int. J. Mass Spectrom. Ion Phys. 34, 159 (1980).
8   C.T. Rettner and J.H. Brophy, Chem. Phys. 56, 53 (1981).
9   V.S. Antonov, V.S. Letokhov and A.N. Shibanov, Appl. Phys. 22, 293 (1980).
10 V.S. Antonov, and V.S. Letokhov, Appl. Phys. 24, 89 (1981).
11 D.H. Parker and R.B. Bernstein, J. Phys. Chem. 86, 60 (1982).
12 C. Klimcak and J. Wessel, Appl. Phys. Lett. 37, 138 (1980).
13 T.W. Scott, C.L. Braun and A.C. Albrecht, J. Chem. Phys. 76, 5195 (1982).
14 K. Siomos, G. Kourouklis and L.G. Christophorou, Chem. Phys. Lett. 80, 5041 (1981).
15 V.S. Antonov, V.S. Letokhov and A.N. Shibanov, Appl. Phys. 25, 71(1981).
16 R.J.S. Morrison, B.H. Rockney and E.R. Grant, J. Chem. Phys. 75, 2643 (1981).
17 T.E. Carney and T. Baer, J. Chem. Phys. 75, 4422 (1981).
18 Y. Achiba, K. Sato, K. Shobatake and K. Kimura, J. Chem. Phys. 77, 2709 (1982).
19 J. Silberstein and R.D. Levine, Chem. Phys. Lett. 74, 6 (1980).
20 J. Silberstein and R.D. Levine, J. Chem. Phys. 75, 5735 (1981).
21 F. Rebentrost, A. Ben-Shaul and K.L. Kompa, Chem. Phys. Lett. 77, 394 (1981).

22   F. Rebentrost and A. Ben-Shaul, J. Chem. Phys. 74,
      3255 (1981)
23   W. Dietz, H.J. Neusser, U. Boesl, E.W. Schlag and S.H. Lin,
      Chem. Phys. 66, 105 (1982).
24   J.J. Yang, D.A. Gobeli, R.S. Pandolfi and M.A. El-Sayed,
      (to be published).
25   R.L. Whetten, Ke-Kian Fu, R.S. Tapper and E.R. Grant,
      J. Phys. Chem. (in print).
26   J.T. Meek, R.K. Jones and J.P. Reilly, J. Chem. Phys.
      73, 3503 (1980).
27   J.C. Miller and R.N. Compton, J. Chem. Phys. 75, 2020
      (1981).
28   P. Hering, A.G.M. Maaswinkel and K.L. Kompa,
      Chem. Phys. Lett. 83, 222 (1981).
29   T.E. Carney and T. Baer, J. Chem. Phys. 76, 5963 (1982).
30   D. Proch, D.M. Rider and R.M. Zare, Chem. Phys. Lett. 81,
      430 (1981).
31   U. Boesl, H.J. Neusser and E.W. Schlag, Chem. Phys. Lett.
      87, 1 (1982).
32   R.S. Pandolfi, D.A. Gobeli and M.A. El-Sayed, J. Phys.
      Chem. 85, 1779, (1981).
33   G. Forst, Theory of Unimolecular Reactions, (Academic,
      New York 1973).
34   T. Baer, A.E. DePristo and J.J. Hermans, J. Chem.
      Phys. 76, 5917 (1982).

MULTIPHOTON SELECTIVE EXCITATION: ON THE ROLE OF CHAOTIC DYNAMICS

WITH MANY BASINS OF ATTRACTION [*]

F. T. Arecchi

Istituto Nazionale di Ottica and
Phys. Dept., University, Firenze

ABSTRACT

Considering the multiphoton excitation of a large molecule as a
classical nonlinear dynamics, some speculations are offered on the
possible implications of a chaotic dynamics with many basins of
attraction.

The problem of selective laser excitation of complex molecules
by infrared multiphoton absorption has received much attention
because of its possible applications to chemical engineering on
one side, and of its implications in statistical mechanics on the
other side. I wish to direct that attention to the problem of relaxations in unimolecular reactions.
    Since many reviews are available on this subject [1,2] I shall
assume that the current conjectures on the loss of selectivity
after a few-photon absorption are well known. To be precise, molecular activation is based on RRKM theory (uniform heating of a
molecule even for resonant excitation of one vibrational mode)only
for very rapid intramolecular relaxtions [3]. The theory breaks down
if the energy is supplied to one degree of freedom within a time
shorter than the time scale of intramolecular vibrational transfer
to other modes, thus allowing for selective bond chemistry.
    The non-RRKM chemistry is modeled after the spin-lattice relax-

(*) Work partly supported by Contract CNR-INO

ation theory in NMR [4] or the electric dipole-bath relaxation in quantum optics [5], that is, it is based on a coherent interaction between an IR field and a resonant vibrational mode, disturbed by diagonal and off-diagonal relaxation terms toward a thermal bath made up by all the other degrees of freedom of the molecule. This model implies a single time scale

$$t \ll T_1, T_2$$

where $T_1 \simeq T_2$ are the relaxation times toward a structureless bath having the usual maximum entropy properties.

It has been recently shown [6,7] that chaotic phenomena, occurring on a time scale corresponding to a high frequency spectrum, can be also characterized by a second, much longer, time scale (corresponding to a low frequency component) if the following circumstances are verified: i) the system has at least two attractors; ii) the attractors are near to be destabilized or they have just become unstable; and iii) the system is "open" to external fluctuations, i.e., the presence of white noise is essential to yield jumps between different basins of attraction. The long time scale yields a diverging low frequency spectral power going as $\omega^{-\alpha}$. In our spectra [6] the slopes should all be multiplied by a factor 2, because of a misleading use of the power calibration in the spectrum analyzer. Hence most slopes cluster around $\alpha = 1.2$. The above conditions show that we are in presence of a phenomenon which occurs beyond the usual approach to chaos by either one of the current scenarios [8]. A simple jump between two attractors is not sufficient to explain the phenomenology. In fact experiments [6] show that, when leaving an attractor, the representative point in phase space has a long erratic motion before landing onto another attractor. Usually our spectra extend over less than two decades, at frequencies much less than the driving frequency. For the CO_2 laser [7], we have a $f^{-0.6}$ spectrum between 1 and 100 Hz, while the driving frequency is around 60 KHz.

To understand the above features we have investigated numerically [9] the dynamical equations of the systems of Ref. 6 and 7. However the 1/f region is so narrow that this procedure gives a poor insight. We have preferred to build a recursive map allowing for two independent attractors, namely

$$x_{n+1} = (a - 1)x_n + a x_n^3$$

on the interval $(-1,1)$ disturbed by white noise of variance between 10^{-7} and 10^{-5}. a is set around 3.980006 in order to have two period-three attractors which have become strange and extended over intervals of the order of 10^{-3}. As in the experiments [6,7], we observe $f^{-\alpha}$ regions with $\alpha = 1$ for intermediate noise levels, and $\alpha \to 0$ for high noise and $\alpha \to 2$ for low noise. These 1/f regions extend over 2 decades with slope changes at the two extrema, thus appearing as a suitable superposition of Lorentzians. As in equi- librium systems a 1/f slope is the result of a sum over a large number of Lorentzian curves suitably weighted[10], similarly we formulate the conjecture that in a nonlinear system we have to sum over many stability valleys to extend the 1/f law over many decades.

A speculative application of the above considerations to selec- tive photochemistry is as follows. In multiphoton dissociation or isomerization reactions, consider the molecular potential as the anharmonic potential of a classical dynamical system. The nonlinear evolution under a driving radiation field near to resonance with the linearized eigen frequency (corresponding to the harmonic part of the potential) may allow for two or more independent attractors, as in Ref. 6, 7, 9.

How "classical" is the motion of a many-atom molecule is still an open question[11]. Let me remind however two powerful theorems[12,13] which state that, when coupling two systems (e.g. one an electromagnetic field and the other a molecular system) if one is classical, the other evolves from the ground state to a coherent state, which is very different from an energy eigenstate. This coherent state (of the field, if the driving source is a classical current[12]; of the atom if the source is a classical e.m. field[13]) is a minimum uncertainty state. Hence the natural quantum state of a system driven by a classical source is very well approximated by a classical description. We should expect that a molecule driven by a high intensity IR field reaches a coherent state with a specific algebra depending on the symmetry of the problem. Coherent states for harmonic oscillators were introduced in Ref.12, for N 2-level atoms in Ref. 13, and N-levels atoms in Ref. 14. These References may offer hints on how to work coherent states for particular molecular potentials. What here matters, however, is the fact that the current pictures on classical chaos may be relevant to describe a multiphoton unimolecular process.

Thus, it should be easy to adjust the control parameters (fre- quency and intensity of the IR field) in order to have a many- attractor dynamics, and hence a structured thermal bath made of two, or more, basins of attraction, with individual relaxation times toward each of them, plus much longer jump times. If these longer times occur much above the picosecond range, it should then be easy to observe phenomena which are secular with respect to the current picosecond intramolecular relaxations. For instance, it should be possible the excitations of one, over two, local groups of bonds even with rather long pulses. Such a possibility was already shown by Lin et al. [15] for a non-centrosymmetric molecule as sucrose solid in KBr.

In this short outlook of "chaotic photo-chemistry" I have used terms which may not be familiar to the audience. I find thus convenient to add a tutorial Appendix on very recent investigations of these phenomena.

APPENDIX

Nonlinear dynamical systems and turbulence

Let us consider a dynamical system, ruled by the equation

$$\dot{X} = \vec{F}(X; \mu) \tag{1}$$

where X is an n-dimensional vector, $\vec{F}$ a nonlinear function and μ an m-dimensional control parameter.

We study the equilibrium solutions

$$F(X; \mu) = 0 \tag{2}$$

for different μ. For a critical value of μ, a stationary solution may switch from stable to unstable (bifurcation). An example is the laser threshold, where the branch $x_1 = 0$ becomes unstable, and it appears a new stable branch $x_2 \neq 0$.

We are not interested in the general treatment of bifurcations, but just how a sequence of them may lead a system to *turbulence*, or chaos. We call turbulence the appearance of a continuous power spectrum, which corresponds to unpredictable features (a line power spectrum would correspond to a periodic correlation function).

Before 1963, the Landau-Hopf model [15], based on mode-mode coupling in a fluid due to the nonlinear hydrodynamic equations, hypothized the generation of a large amount of uncommensurate frequencies, which eventually accumulated into a continuum. The hydrodynamic equations, being field equations correspond to an infinite number of coupled equations. In 1963 Lorenz [16] showed that three coupled nonlinear equations were enough to reach chaos. Now, three equations can mimic field equations with drastic cut-offs due to the boundary conditions.

If in eq.(1) we take $X = (x,y,z)$, we have a three-dimensional phase space. After a transient, the trajectories become closed loops (periodic or stable attractors) corresponding to power

spectra made of discrete lines. But for crucial values of the
control parameters $\vec{\mu}$, one has a *strange* attractor, that is, an
intricate loop which never appears to close on itself, and correspondingly a continuous spectrum.

The Lorenz equations, with parameter values suitable to give a
strange attractor, are given by

$$\dot{x} = - 10\ x + 10\ y$$
$$\dot{y} = -\ x\ z + 28\ x - y$$
$$\dot{z} = x\ y - 8/3\ z \tag{3}$$

Notice that Lorenz chaos is a *deterministic* one, because we do *not*
have noise sources.

Equivalent to the Lorenz model is a system of two 1st order
equations (or one 2nd order equation) plus an external modulation.
An example is the driven Duffing oscillator[17,6] ruled by

$$\ddot{x} + \gamma\dot{x} + \omega_o^2 x - \beta x^3 = A\cos\omega t \tag{4}$$

Equation (4) is equivalent to 3 coupled equations

$$\dot{x} = y$$
$$\dot{y} = -\gamma y - \omega_o^2 x + \beta x^3 + A\cos z$$
$$\dot{z} = \omega \tag{4'}$$

The potential corresponds to a single minimum. For different control parameters $\vec{\mu}$ (either modulation amplitude A or frequency ω)
it may give a sequence of subharmonic bifurcations leading eventually to chaos as shown in fig. 1.

The sequence of subharmonic bifurcations corresponds to the
successive appearance of period T = $2\pi/\omega$, 2 T, 4 T... 2^nT in
the output. If we call μ_n the value of the parameter at which
the period 2^nT appears then the following relation is verified

$$\delta = \frac{\mu_{n+1} - \mu_n}{\mu_{n+2} - \mu_{n+1}} \xrightarrow{n \gg 1} 4.669...$$

This number has been shown by Feigenbaum[18] to be universal.

If, in the space $(\dot{x}, x, t)$ one considers the discrete transformation from the point $(\dot{x}, x)$ at time t to the point $(\dot{x}, x)$ at
time t + T after integration of equation (4) over that interval
one has the discrete mapping

$$\vec{x}_{t+T} = f(\vec{x}_t) \tag{5}$$

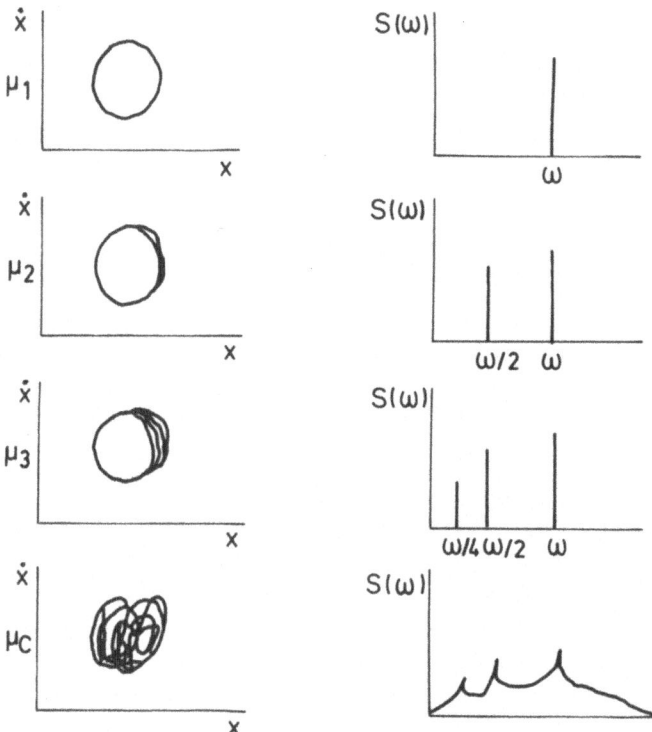

Figure 1. Phase space plots ($\dot{x}$, x) and power spectra S (ω) for
different control parameters.

$$x_{t+2T} = f(x_{t+T}) = f(f(x_t)) = f^{(2)}(x_t)$$

and a fixed point is ruled by the equation

$$x_{t+T} = x_t = f(x_t)$$

Of course, a fixed point in the map *does not* mean a single equilibrium point, like in the differential equation, but a limit cycle of period T, whereby the position in phase space goes onto itself at each T. A period 2T would appear as a solution of the equation

$$x_{t+2T} = x_t = f^{(2)}(x_t)$$

and so on. Feigenbaum has evaluated his δ value by showing that, in such cascades of period doublings, the local structure of the attractor is reproduced at a rescaled size in successive bifurcations, with the rescaling parameter being a universal constant.

So far we have sketched how, by varying the control parameter, a system ruled by equation (4) or (5) undergoes a sequence of bifurcations eventually leading to chaos.

Noise is *not* essential (deterministic chaos), but if we add it, the number of subharmonic bifurcations before chaos become smaller and smaller. This can be put in terms of a scaling law where the variance of the external noise appears somewhat as a modification of the control parameter. Let us now change the sign of the potential, getting two stable valleys (fig. 3). This is equivalent to the new Duffing equation:

$$\ddot{x} + \gamma \dot{x} - \omega_o^2 x + \beta x^3 = A \cos \omega t \qquad (4)''$$

Depending on the initial conditions, we have *two* independent attractors. Let us increase μ until both attractors become *strange*. Now, addition of a random noise may trigger jumps from one to the other. These jumps give a low frequency divergence in the power spectrum[6]. These jumps may be considered as a *superchaos* insofar as they couple two strange attractors, otherwise independent. Here presence of random noise is essential.

Fig. 4b shows the appearance of the low frequency, or $1/f$ divergence. A similar effect was observed in a Q-modulated CO_2 laser.[7]

Fig. 4 shows bistability, that is, the simultaneous coexistence of two attractors corresponding respectively to $f/4$ and $f/3$ subharmonic. Increasing the modulation depth m, the attractors become

Such a correspondence is illustrated in fig. 2. It is called
stroboscopic, or Poincaré, map. The difference equation (5) is
fully equivalent to the differential equation (4). Of course,
having performed a time integral, one has reduced the three-dimensional recurrence (5). In many cases of physical interest the points
of the Poincaré map cluster over an almost one-dimensional manifold.

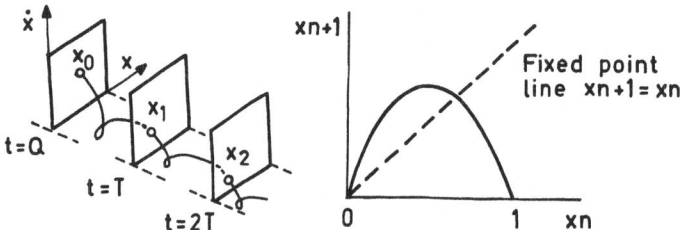

Figure 2.   Trajectory in 3-D phase space (solid line) and direct
            mapping $x_t \to x_{t+T}$ (dashed line). Example of quadratic map.

Since the interesting bifurcations are associated with a change of
slope, the one-dimensional map can be studied around a maximum.
Thus a quadratic map as

$$x_{n+1} = \lambda x_n (1 - x_n) \tag{6}$$

is sufficient to display most of the features of equation (3) and
the role of μ is here taken by λ. One can develop a straightforward set of transformations using discrete maps[19]. For instance the second iterate is

strange and correspondingly one observes the spectral divergence
as in fig. 4c. Thus associating low frequency divergence with
jumps between two attractors (multistable situations) seems a
successful conjecture.

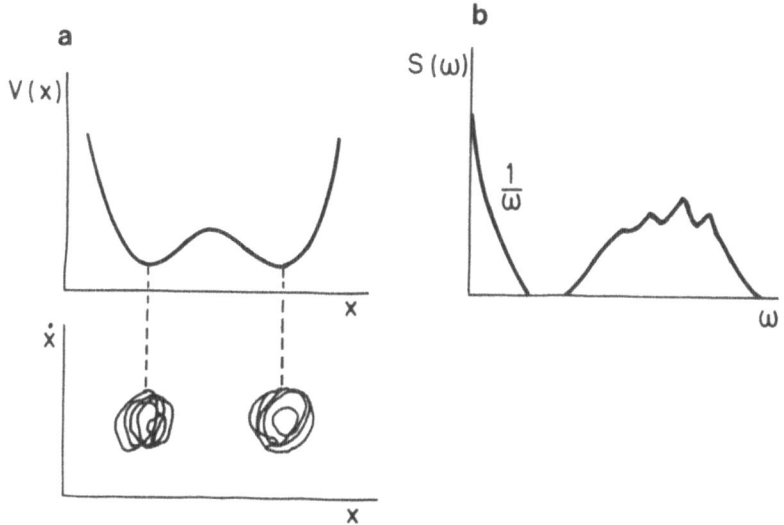

Figure 3. a) Bistable potential V(x) (change of sign in an-harmonic
              force of eq. (4)). Coexistence of two strange attract-
              tors in (ẋ,x) with possible mutual jumps induced by
              external noise (line with arrows).

           b) Birth of 1/ω branch in power spectrum associated with
              the above jumps.

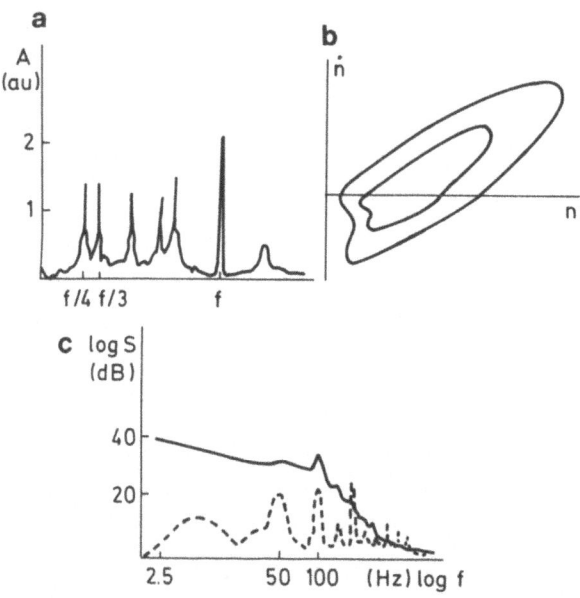

Figure 4. Bistability and 1/f noise in a CO_2 laser with loss
modulation

        a-b) coexistence of two attractors (period 3 and 4 re-
             spectively). The two superposed spectra correspond
             to two starts with different initial conditions.
          c) comparison between the low frequency cut-off
             (dashed line) when the two attractors are stable
             and the low frequency divergence (solid line,
             slope $\alpha=0.6$) when the two attractors are strange.

REFERENCES

1. N. Bloembergen and E. Yablonovitch, Physics Today, May 1978,
    page 83
2. A. H. Zewail, Physics Today, November 1980, page 27
    W. Fuss and K. L. Kompa, Progr.Quant.El. 7:117 (1981)
3. P. J. Robinson, K. A. Holbrook, Unimolecular Reactions, Wiley-
    Interscience,1972
4. A. Abragam, Nuclear Magnetism, Oxford 1961
5. H. Haken and M. Sargent and M. O. Scully, in:"Laser Handbook",
    ed.by F. T. Arecchi and E. Schulz Du Bois, North Holland 1972,
    vol. 1
6. F. T. Arecchi and F. Lisi, Phys.Rev.Lett. 49:94 (1982)
7. F. T. Arecchi, R. Meucci, G. Puccioni and J. Tredicce, Phys.
    Rev.Lett. 49:1217 (1982)
8. J. P. Eckman, Rev.Mod.Phys. 53:643 (1981)
9. F. T. Arecchi, R. Badii and A. Politi (to be published)
10. R.F. Voss and J. Clarke, Phys.Rev. B13:556 (1976)
11. W. E. Lamb, in: "Laser Spectroscopy IV", ed. by H. Walther and
    K. W. Rothe, Springer 1979
12. F. Bloch and A. Nordsieck, Phys.Rev. 52:54 (1937)
    R. J. Glauber, Phys.Rev. 131:2766 (1963)
13. F. T. Arecchi, E.Courtens, R. Gilmore and H. Thomas, Phys.Rev.
    A6:2211 (1972)
14. F. T. Arecchi, R. Gilmore and D. M. Kim, Lett.N.Cimento 6:219
    (1973)
15. L. Landau and E. Lifshitz, "Statistical Physics", Pergamon
    Press, 1958
16. E. Lorenz, Jour.Atmos.Sc., 20:130-141 (1963)
17. J.P. Crutchfield and B. A. Huberman, Phys.Rev.Lett. 43:1743
18. M. J. Feigenbaum, J.Stat.Phys. 19:25 (1978) ; 21:669 (1979)
19. P. Collet and J. P. Eckmann, "Iterated Maps, Birkhäuser (1980)
20. C. T. Lin, J. B. Valim and C. A. Bertran, in: "Lasers and
    Applications", ed. by W. O. N. Guimares et al., Springer
    Series in Optical Sciences, vol. 26, 1981

TWO APPLICATIONS OF LASERS:
I. MULTIPHOTON EXCITATION OF CHEMICAL REACTIONS
II. MODE SPECIFIC EXCITATION OF BIMOLECULAR REACTIONS

George C. Pimentel

Chemistry Department
University of California
Berkeley, California 94720

The special properties of lasers have made possible investigation of a variety of chemical phenomena not accessible with conventional light sources. One of these special properties is high intensity (high photon flux) and another is spectral purity (narrow frequency range). I will tell about two experiments being conducted in my laboratory that demonstrate how these characteristics reveal chemical behaviors not even anticipated before lasers came on the scene. The first of these is called "multiphoton excitation" and the second "mode specific behavior."

MULTIPHOTON EXCITATION OF CHEMICAL REACTIONS

Large carbon dioxide lasers are now commercially available that will deliver infrared light pulses of 75 to 150 nanosecond duration with single pulse energy of several joules. At a typical laser wavelength, 940 cm^{-1}, a three joule pulse contains about 10^{20} photons. Delivered in 150 nsec, this corresponds to a flux of the order of 10^{27} photons/sec. With such incredible light intensities, new photolytic behaviors have been discovered.   One of these is called multiphoton vibrational excitation. As described by two of its discoverers, R.V. Ambartzumian and V.S. Letokov[1], a CO_2 laser can be used to cause a single gaseous molecule to absorb many tens of photons in a time short compared to kinetic collision times. In the early manifestations of multiphoton excitation, molecules were brought up to internal energies sufficient to rupture chemical bonds. This unexpected behavior has been the subject of vigorous investigation by numerous researchers.[2] I would like to present one such study, conducted by Dr. Robert A. Stachnik in my laboratory at Berkeley.[3]

Because multiphoton excitation funnels energy into the molecule through its vibrational degrees of freedom, it produces molecules with an energy distribution quite different from those produced through thermal or electronic excitation. Whether distinctive chemistry is associated with such a distinctive energy distribution remains, as yet, unclear. Dr. Stachnik assembled experimental equipment suited to explore this question. I will discuss one of his applications of this equipment, i.e. to the molecule tri-fluoroethene, F_2CCHF.

A grating-controlled Lumonics 103 TEA CO_2 laser was tuned to a laser wavelength absorbed by F_2CCHF and a modified Perkin-Elmer 270 double focus mass spectrometer was used to analyze the products of any chemical reactions that could be activated. With a repetition rate of less than one pulse per second, the laser provided three joule pulses with a nominal duration of 150 nsec. These pulses were brought to a diffuse focus about 3 mm from the entry orifice to the mass spectrometer. The gas under study, either pure F_2CCHF or a mixture with a deliberately added reactant, was passed through this focal region under slow flow conditions to guarantee that the mass spectrometer would record the products associated with a single pulse. The total pressure, 0.1 to 0.5 torr, was selected to permit vibrationally excited molecules about 10^3 collisions before entering the mass spectrometer. Thus the arrangement permits the investigation of the chemistry of molecules produced with excess energy in their vibrational modes.

To support these experiments, infrared spectra of the products of 30 to 300 pulses were recorded with a FTIR Spectrometer, thus removing some of the ambiguity intrinsic to mass spectrometry.

In a typical experiment with 0.25 torr of pure F_2CCHF and without a laser pulse, the mass spectrum shows the complex fragmentation pattern of the parent molecule as well as a variety of ion peaks of higher mass than the parent peak at 82 mass units. These higher peaks show that in the ionization region of the mass spectrometer, initially formed ions undergo polymerization, elimination and hydrogen abstraction reactions. This background spectrum inhibits but does not prevent detection of laser-induced chemistry and it forewarns that the mass spectrum must be interpreted with care to distinguish the desired reactions that occur before passage through the entry orifice from those that take place in the ionization chamber.

Only two experiments will be described to illustrate the potentialities of the technique, the laser excitation of pure F_2CCHF and of F_2CCHF mixed in a 1/3 mixture with ethylene. The mass spectrum of laser-excited pure F_2CCHF showed the new mass peaks listed in the first column of Table I. The mass peak 94 shows the greatest enhancement when the laser is turned on.

Table I

Mass Peaks Produced by Laser Irradiation at 940.5 cm^{-1} of
Pure F_2CCHF and of $1/3=F_2CCHF/C_2H_4$ (0.5 torr)

	Pure F_2CCHF			F_2CCHF/C_2H_4	
Mass	Growth [a]	Identity	Mass	Growth	Identity
94	179	$C_3F_3H^+$	40	143	$C_3H_4^+$
93	22	$C_3F_3^+$	39	$\sim$50	$C_3H_3^+$
75	139	$C_3F_2H^+$	38	89	$C_3H_2^+$
74	19	$C_3F_2^+$	37	38	C_3H^+
100	55	$C_2F_4^+$	100	34	$C_2F_4^+$
106	18	$C_4F_3H^+$			
86	47	$C_4F_2^+$			

a. Number given times 10^3 is the number of counts recorded over
   background.

This mass corresponds to the ion formula $C_3F_3H^+$, a parent F_2CCHF
molecule plus a carbon atom. The next three mass peaks listed,
93, 75, and 74, can be plausibly related to $C_3F_3H^+$ through loss
of an H atom, F atom, and HF molecule, respectively. The next peak,
mass 100, is that of $C_2F_4^+$. These five peaks, which include the
three that show greatest growth, are most informative when compared
to the mass spectrum of the $1/3=F_2CCHF/C_2H_4$ experiment.

The fourth column lists the masses that grow when a three-fold
excess of C_2H_4 is added while the total pressure is held constant.
Now the mass 94 peak grows but ten-fold less and the most intense
peak is mass 40. This mass corresponds to the collision partner,
C_2H_4, plus a carbon atom. The next three mass peaks listed are
counterparts to those in the first column. The fifth mass peak
again corresponds to C_2F_4 but reduced only by 0.6 from the growth
observed with pure F_2CCHF. These results and others that cannot
be presented because of time limitations show that the following
chemical reactions are indicated.

$$F_2CCHF + nh\nu \quad \rightarrow \quad F_2CC + HF \quad\quad (1)$$

$$F_2CC + F_2CCHF \quad \rightarrow \quad F_2CCCHF + CF_2 \quad\quad (2a)$$

$$F_2CC + H_2CCH_2 \qquad \rightarrow \qquad H_2CCCH_2 + CF_2 \qquad (2b)$$
$$CF_2 + CF_2 \qquad \rightarrow \qquad C_2F_4 \qquad\qquad\qquad (3)$$

Reaction (1) shows the vibrationally induced $\alpha\alpha$ elimination of HF from the difluorovinylidine[4] molecule. This reaction is endothermic by about 80 kcal/mole[4] so it requires that each molecule absorb at least 30 quanta of laser radiation. It is possible that $\alpha\beta$ elimination also takes place since the difluoroacetylene mass peak 62 is obscured by a heavy background peak. The $\alpha\beta$ elimination is only about 38 kcal/mole[5] endothermic but there is undoubtedly a large activation energy associated with this process that permits reaction (1) to compete kinetically.

Because of the endothermicity of reaction (1), the F_2CC molecule will be produced with relatively little internal excitation. Hence heavy atom migration to form the relatively stable difluoroacetylene is not facile. Instead, we feel that the subsequent chemistry is that of difluorovinylidine. Unexpectedly, this molecule seems to act as a carbon insertion reagent with olefins to produce the appropriate allene and CF_2 molecule. Then reaction (3) consumes the CF_2 molecules to product C_2F_4.

It is necessary, of course, to verify that the mass peaks shown in Table I are not all due to fragmentation of much larger molecules formed through polymerization in the ionization chamber. The infrared spectra of static samples exposed to many laser pulses provide such proof. The infrared spectrum of F_2CCHF exposed to 100 laser pulses[6] shows growth of absorptions at 2030, 1088, and 1247 cm^{-1}, all due to trifluoroallene[6] and at 1340 and 1187 cm^{-1} due to C_2F_4. The 1/3 mixture with C_2H_4 has in its product spectrum the well-resolved rotational fine[4] structure of the allene ν_{10} absorption near 840 cm^{-1} as well as the absorptions of C_2F_4 and $F_2C=C=CHF$.

The data reveal other processes at work, of course. For example, the mass peak 106, $C_4F_3H^+$, could be due to a second carbon insertion of F_2CC into product trifluoroallene. On the other hand, in the C_2H_4 experiment, the mass peak 70, $C_4H_3F^+$ might be evidence for stabilization of the addition product of difluorovinylidine to ethene to produce $C_4H_4F_2$. Similarly, the mass peak 58, $C_3H_3F^+$, might be evidence for addition of CF_2 across the ethene double bond to produce $C_3H_4F_2$. Further information is to be derived from the infrared spectra. For our purposes, however, the data indicate that an important primary process in the multiphoton vibrational excitation of trifluoroethene is the $\alpha\alpha$ elimination of HF and that the resulting difluorovinylidine acts as a carbon insertion agent to produce allenes. Plainly interesting and unusual chemistry can be excited through multiphoton excitation and studied by the techniques just described.

MODE SPECIFIC EXCITATION OF BIMOLECULAR REACTIONS

The availability of highly monochromatic laser light sources
at a wide range of frequencies makes possible the selective excitation of molecules into particular and known states. This raises
the possibility of examining the chemical reactivity of such selectively prepared molecules. Such an experiment is called "mode
specific excitation" and it conjures an alluring vision of the
systematic study of the relative importance of each degree of
freedom in chemical reactivity.

These widely held hopes have resulted in many imaginative
experiments but, for polyatomic molecules, relatively little
success thus far. The difficulty seems to be that intramolecular
energy movement is faster than had been anticipated so that the
molecular memory of its selective excitation is too short-lived
to reflect in its bimolecular chemistry. A part of this rapid
intramolecular energy redistribution is associated with the high
energy level densities that characterize even relatively low-lying
states of polyatomic molecules.

Partly with this level density issue in mind, we have attempted
to initiate mode-selective chemistry of reactive molecular pairs
embedded in a cryogenic matrix such as solid argon or solid nitrogen
at 12°K. I would like to describe briefly some experiments of this
type conducted by Dr. Heinz Frei on the ethylene-fluorine reaction[7,8]
and by Mr. Arne Knudsen on the allene-fluorine reaction.[9]

A typical experiment begins using conventional double deposition
matrix isolation techniques. Simultaneous deposition of a nitrogen/
fluorine mixture and a nitrogen/olefin mixture prevents prereaction
and gives a matrix containing isolated olefin, isolated fluorine,
and olefin/fluorine aggregates in relative amounts determined by
the matrix concentrations and the mode of deposition. With olefin/
fluorine/nitrogen mole ratios of 1/1/100, approximately 10% of the
olefins will be found with a fluorine nearest neighbor, a "reactive
pair." An FTIR spectrometer can, in many cases, resolve and dif-
ferentiate between the absorptions of an olefin molecule isolated
in the matrix from the slightly shifted absorption of the olefin
when it has an F_2 nearest neighbor. This implies that a tuned
laser can be selected whose frequency is absorbed preferentially
by these potentially "reactive pairs." If reactivity results from
a period of irradiation, it is manifested by loss of reactive pair
absorption and by growth of absorptions due to products. Again an
FTIR instrument is well suited to display such spectral changes in
its ability to present difference spectra.

Dr. Frei's experiments with the ethylene fluorine system are
most advanced.[7,8] Three laser systems were used for selective vibrational excitation, a grating-controlled CO_2 laser for excitation

in the range 915-1092 cm^{-1}, a grating-controlled CO laser for the range 1550-2000 cm^{-1}, and an F-Center continuously tunable laser (Burleigh, model FCL-10) for the range 2950-4350 cm^{-1}. The first two lasers produce line-tunable continuous power levels up to one watt per square centimeter. This is about the maximum power that can be used without thermal disturbance of the cryogenic matrix. The F-Center laser delivers much lower power, in the range 5-80 milliwatts per square centimeter but is continuously tunable.

Six well characterized vibrational modes of ethylene were excited with frequencies ranging from 953 cm^{-1} (ν_7) to 4209 cm^{-1} ($\nu_3+2\nu_7+\nu_{10}$). Over this modest range of excitation energy, the quantum yield for reaction changed by five orders of magnitude. The chemistry was that first deduced by Hauge and coworkers:[10]

$$C_2H_4 + F_2 \quad \rightarrow \quad \underline{trans}, \underline{gauche}\text{-}H_2FCCH_2F \quad (1)$$

$$H_2FCCH_2F \quad \rightarrow \quad HF + H_2CCHF \quad (2)$$

Infrared absorptions due to both products, 1,2-difluoroethane and vinyl fluoride, grow with first order kinetics which, through curve-fitting, can be converted to a quantum yield. Change of the laser power level shows that single photon excitation is involved.[8]

Table II lists the measured quantum yields.[8] No products could be detected after prolonged irradiation at 953 cm^{-1}; our sensitivity level for product detections provides the upper limit on this quantum yield. This absorption excites the fundamental of the b_{1u} out-of-plane hydrogen bend, ν_7, which is a plausible reaction coordinate for a four center addition. Reaction is observed, however, with excitation at 1896 cm^{-1} where the combination $\nu_7+\nu_8$ absorbs. With about double the energy, this mode causes reaction with quantum yield at least two orders of magnitude higher than at 953 cm^{-1}. Looking at all of the quantum yields, it is plain that there is an overall tendency for quantum yield to rise dramatically with photon energy.

These data and similar results for the three dideutereothylenes point to a model. Upon excitation of a particular vibrational eigenstate, the molecule has choice between reaction with its fluorine neighbor or dropping to some lower vibrational state of lower energy while transferring the energy difference into the matrix crystal lattice vibrations, the so-called "phonon modes." Since the ethylene rotational degrees of freedom have been "frozen out", its energy levels are rather sparse. This means that each such relaxation step must transfer a considerable amount of energy from internal C_2H_4 vibrational energy into lattice heat. Since the matrix interacts quite weakly with guest impurities, such energy transfer is relatively improbable. It is reasonable to expect, further, that the larger the energy increment to be transferred to

Table II

Quantum Yields for Selective Excitation of Ethylene-Fluorine Pairs
in Solid Nitrogen at 12°K[8]

C_2H_4 Mode	ν (cm^{-1})	ϕ_ν
ν_7	953	$3 \cdot 10^{-6}$
$\nu_7 + \nu_8$	1896	$6.6 \cdot 10^{-4}$
ν_{11}	2989	$2.3 \cdot 10^{-2}$
$\nu_2 + \nu_{12}$	3076	$7.0 \cdot 10^{-2}$
$2\nu_9$	3105	$4.3 \cdot 10^{-2}$
$\nu_3 + 2\nu_7 + \nu_{10}$	4209	$3.1 \cdot 10^{-1}$

the lattice, the slower will be the relaxation. This leads to the
expectation that relaxation will tend to take place in relatively
small steps. Such a cascading process implies, in turn, that the
quantum yield will tend to rise with photon energy simply because
the molecule cascades through more and more potentially reactive
vibrational states as it loses its energy.

That is not, however, evidence for mode selectivity. If a
particular vibrational mode is excited which has a specially
favorable propensity for reaction, it will display this by a
specially high increase in quantum yield compared to the average
trend. Thus mode specificity is revealed by looking at the changes
in quantum yield as photon energy increases. This is most
strikingly displayed by the quantum yield at 3076 cm^{-1} which is a
factor of 1.6 higher than that observed at 3105 cm^{-1} and a factor
of 3 above that at 2989 cm^{-1}.

Mr. Arne Knudsen is presently studying the allene-fluorine
reaction with comparable results.[9] The chemistry is somewhat more
complicated because the expected (and observed) addition product,
2,3-difluoropropene, can eliminate HF in two ways. One elimination
route produces 3-fluoropropyne (propargyl fluoride) and the other
produces fluoroallene. Since the difluoropropene has two rotamer
conformations, cis- and gauche-, there are four product molecules
contributing to the product spectrum. Nevertheless it has been
possible to identify all four products, aided considerably by the
fact that the two rotamer conformers can be reversibly interconverted
by tuning one of the lasers to an absorption of one or the other
molecule.

The two elimination channels give this study more dimension
than the ethylene case. For example, choice of the matrix affects
the relative yield of the different products. In xenon almost no

propargyl fluoride is detected and fluoroallene is a dominant product whereas these two products are more nearly equal in nitrogen matrix.

Returning to the theme of mode selectivity, only four quantum yields have yet been measured. Again they display a high dependence on total photon energy, ranging from around 10^{-4} at 1679 cm^{-1} ($2\nu_{10}$, B_2) to 0.2 at 3076 cm^{-1} (ν_8, e). Again, however, particularly high propensity for reaction seems to be associated with excitation of $2\nu_9$ at 1999 cm^{-1} relative to excitation of ν_6 at 1953 cm^{-1}.

Thus our cryogenic experiments seem to demonstrate mode selective behavior in the vibrational excitation of fluorine-olefin addition reactions. There remains the question of why this set of experimental conditions is apparently suited to investigate such behavior. We feel that one of the factors at work is the quenching of rotational degrees of freedom to reduce the density of levels. Substituted for these are, of course, the lattice phonon modes but it is the essence of the matrix isolation technique that a guest molecule interacts quite weakly with the lattice. In this particular study, the interaction must be such that it provides a relaxation route that is competitive with reaction. Some such competition would seem to be a necessary feature of mode-selective experiments since it permits the sensing of different reaction propensities. If this is a correct inference, then it suggests that a better understanding of these matrix relaxation steps will help as we attempt to generalize by moving on to other vibrationally-induced reactions in cryogenic environments. In any event, these studies, like those described earlier on multiphoton excitation, are made possible by the existence of laser light sources. Because of these new possibilities, we can look for rapid advances in our understanding of how chemical reactions occur.

BIBLIOGRAPHY

1. R.V. Ambartzumian and V.S. Letokov, Chemical and Biochemical Applications of Lasers, Vol. III, ed. C.B. Moore, Academic Press, N.Y. (1977).
2. e.g., see P.A. Schulz, Aa. S. Sudbo, D.J. Krajnovich, H.S. Kwok, Y.R. Shen, and Y.-T. Lee, Ann. Rev. Phys. Chem. 30, 379 (1979).
3. Robert A. Stachnik, Ph.D. Dissertation, University of California, Berkeley (1981).
4. M. Frisch, R. Krishman, J.A. Pople, and P. Von Schleyer, Chem. Phys. Letters, 81, 421 (1981).
5. O.P. Strauz, R.J. Norstrum, D. Salahub, R. Gosavi, H.E. Gunning, and I.G. Csizmadia, J. Am. Chem. Soc. 92, 6395 (1970).
6. J.Y. Durig, Y.S. Li, J.D. Witt, A.P. Zens, and P.d. Ellis, Spectrochim. Acta, 33A, 529 (1977).
7. H. Frei, L. Fredin, and G.C. Pimentel, J. Chem. Phys. 74, 397 (1981).

8.   H. Frei and G.C. Pimentel, J. Chem. Phys. <u>77</u>, xxxx(1982).

9.   A. Knudsen and G.C. Pimentel, J. Chem. Phys. <u>78</u>, xxx(1983).

10. R. Hauge, S. Gransden, J. Wang, and J.L. Margrave, J. Am. Chem. Soc. <u>101</u>, 6950 (1979).

STATE-TO-STATE CHEMICAL KINETICS STUDIED WITH

LASER-INDUCED FLUORESCENCE

J. Wanner

Max-Planck-Institut
für Quantenoptik
8046 Garching, Fed. Rep. of Germany

ABSTRACT

The laser-induced fluorescence (LIF) method has developed into
one of the most powerful diagnostic techniques in molecular reaction
dynamics. This lecture is intended to give an overview of the state
of the art of the LIF method. In addition to the oral presentation
at the San Miniato meeting where LIF was introduced by choosing
examples of chemical reactions now being studied at our laboratory,
an attempt is made here to give a more comprehensive summary. In
particular, this lecture is intended to provide a key to the
literature of the rapidly growing field of state specific investigations of chemical reactions.

INTRODUCTION

During the last two decades chemical kinetics has progressed to
a stage where it has become possible to follow a chemical reaction out
of selected reagent quantum states into final product states. In
contrast to classical thermal rate kinetics, these experiments give
an insight into the detailed dynamics. The experimental techniques
now available in state-to-state kinetics are summarized in Fig. 1
for a simple triatomic reaction $A + BC \rightarrow AB + C$. The reagent molecule
BC can be state selectively prepared by the methods listed on the
upper left of Fig.1 in vibrational-rotational quantum states v, J
of the electronic state E, with a preferred orientation m. Since
state-to-state experiments are preferentially carried out under
molecular beam conditions, the reaction proceeds at a defined
collision energy E_c. The formation of the AB molecule in the

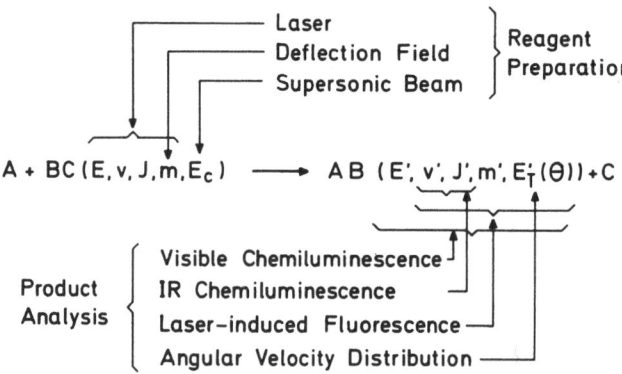

Fig. 1.   Techniques of state-to-state kinetics as applied to
          reagent preparation and product state analysis for a
          simple triatomic reaction.

corresponding product quantum states denoted by a prime can be
measured by the methods listed on the lower right of Fig. 1.
Among diagnostic techniques the laser-induced fluorescence (LIF)
method plays a major role in these investigations. An attempt is
made in the following to give a short review of the potential of
this spectroscopic technique for product state analysis and its
application to the kinetics of state selected species.

1.   INTERNAL PRODUCT STATE ANALYSIS

A general description of the tunable laser fluorescence method
for product state analysis is given by Dagdigian and Zare [1]. This
diagnostic technique requires a low lying stable electronic state
of the product and a tunable laser to probe the nascent distribution.
This way an excitation spectrum is obtained which yields the product
state distribution if the spectroscopic assignment can be made and
if the transition moments are known [1,2]. The state specific product
information is obtained with a high sensitivity. Up to 10 molecules/cm^3
in a single vibrational rotational level have been detected [6].

In the early years LIF was applied to reactions forming group II
and III oxides and halides and OH radicals. A complete list of reactions studied up to January 1977 can be found in the extensive review
of Kinsey [2]. The development from this time up to the present is
summarized in the following. The formation dynamics of further

oxides and halides belonging to Group II and III elements have been investigated [3,4,5] . Vibrational-rotational analysis of iodine-monofluoride could be accomplished in interhalogen reactions of the type F + IX → IF + X (X = Cl, Br, I, R polyatomic) [6] using c.w. dye lasers for product detection. Recently LIF was success-fully applied to study the formation of a polyatomic radical, vinyl oxide, CH_2CO, in the reactions of $O(^3P)$ with olefins [7]. The avail-ability of excimer lasers has resulted in a very recent analysis of the vibrational product state distribution of CO formed in the reac-tion O + CS → CO + S by exciting CO (A → X) fluorescence [8].

As a new frontier, progress can be seen in the application of LIF to ion-molecule dynamics. Product state analysis of the OH radi-cal formed in the reaction O^- + HF → OH + F^- is being investigated at present in a flowing afterglow experiment [9]. Furthermore, there are hopes that the analysis of molecular ions, e.g. formed by reac-tion in the ion cyclotron resonance cell will become feasible [10]. Reaction products to which LIF could be applied so far are listed in Table 1.

Table 1. Reaction products for which LIF internal product
          state analysis was carried out

MO	M	= Mg, Ca, Sr, Ba, Sc, Y, La, Al
MF	M	= Ca, Sr, Ba, Yb
MCl	M	= Sr, Ba
MI	M	= Ba

OH
IF
CH_2CHO
CO

There is the case of triatomic reactions which yield molecular products AB not having a stable, low-lying electronic state suitable for LIF product state analysis. If these reactions are sufficiently exoergic, dynamical information can be obtained from the amount of the electronic excitation of the atom C, which is normaly determined from the visible chemiluminescence. However, LIF can be employed to measure detailed rate constants for the formation of very low-lying atom states. The combination of chemiluminescence and laser-induced fluorescence was successfully demonstrated for the reaction O + Cs_2 → CsO + Cs [11].

2.   ANGULAR VELOCITY DISTRIBUTION MEASUREMENTS

It had already been noted by Dagdigian and Zare that the LIF technique affords the potential for determining product angular velocity distributions [1]. Earlier experiments of the Zare group

for the reactions Ba + LiCl (KCl), Ca + NaCl, Ba + CF$_3$I were carried
out by separating the laser probing region from the reaction zone.
These studies, which demonstrated the feasibility of obtaining this
type of information are summarized in Ref.[12], the information was
limited, however, due to the signal-to-noise ratio. An alternative
elegant technique based on the measurement of the Doppler profile
of a single rotational line has been developed by Kinsey et al.
This concept known as Fourier transform Doppler spectroscopy was
successfully applied to a study of the reaction H + NO$_2$ → OH + NO [13].

3.   PRODUCT ANGULAR MOMENTUM POLARIZATION

The variation of the plane of polarization of the dye laser
offers the opportunity to probe the spatial orientation of reaction
products. A general theory of the measurement of the angular momentum polarization by laser-induced fluorescence has been developed
by Case, McClelland and Herschbach [14]. So far, angular momentum
polarization of BaF determined with laser-induced fluorescence has
been reported in a preliminary investigation of the reaction
Ba + HF → BaF + H by Perry et al. [15].

4.   LIF DIAGNOSTICS IN STATE-TO-STATE EXPERIMENTS

An excellent introduction into these advanced studies has
recently been given by Zare [16]. Unfortunately, there is at present
only a limited number of chemical reactions which could be investigated "truly" state-to-state in a rigorous fashion according to
the scheme in Fig. 1. The most thoroughly studied system is the
endoergic/exoergic pair

$$
\begin{array}{lll}
\text{Sr} & & \text{SrF} + \text{H} \\
& + \text{HF} \quad \rightarrow & \\
\text{Ba} & & \text{BaF} + \text{H.}
\end{array}
$$

LIF internal product state analysis of SrF and BaF allows determination of the effect of selective reagent excitation on the dynamics
of these reactions. In a series of investigations carried out in
different laboratories the reagents were prepared as follows.

a)   Control of translational energy using a supersonic seeded HF
     nozzle beam [17]
b)   Vibrational-rotational state preparation v $\leqq$ 2, employing
     a chemical laser [18]
c)   Metastable Sr*(3P$_1$) atoms were produced by excitaton with a
     dye laser [19].
d)   The effect of selected orientation of the HF reagent molecule
     on the dynamics of the Sr + HF (v = 1, J = 1) reaction was
     studied by rotating the plane of polarization of the HF laser [20].

5.   SUMMARY

     The laser-induced fluorescence technique has been widely ex-
ploited to investigate the energy partitioning in numerous elementary
chemical reactions. A few of them have been studied under "true"
state-to-state conditions. Further progress in this area will largely
depend on the improvement of suitable tunable laser sources. The
probing of energy distributions only partially utilizes the intrinsic
potential of this spectroscopic technique. It has been demonstrated
that besides the scalar quantity of energy distribution, vector
quantities such as the angular velocity distribution and the product
angular momentum polarization can be obtained. Progress certainly will
be seen in this area in the near future.

REFERENCES

1   R.N. Zare, P.J. Dagdigian, Science 185, 739 (1974)

2   J.L. Kinsey, Ann. Rev. Phys. Chem. 28, 349 (1977)

3   K. Liu, J.M. Parson, J. Chem. Phys. 67, 1814 (1977);
     68, 1794, (1978)

4   R. Dirscherl, H.U. Lee, J. Chem. Phys. 73, 3831 (1980)

5   P.J. Dagdigian, J. Chem. Phys. 76, 5375 (1982)

6a   L. Stein, J. Wanner, H. Walther, J. Chem. Phys. 72, 1128 (1980)
 b   T. Trickl, J. Wanner, J. Chem. Phys. 78, 6091 (1983)

7   K. Kleinermanns, A.C. Luntz, J. Phys. Chem. 85, 1966 (1981)

8   C.G. Atkins, G. Hancock, oral presentation, 12th Int. Quantum
     Electronics Conference, Munich 1982

9   C.E. Hamilton, private communication this meeting

10a R. Marx, plenary talk, 7th Int. Symposion on Gas Kinetics,
     Göttingen, August 1972
  b J. Danon, G. Mauclaire, T.R. Govers, R. Marx, J. Chem. Phys.
     76, 1255 (1982)

11   H. Figger, R. Straubinger, H. Walther, J. Chem. Phys. 75, 1,
     (1981)

12   R.N. Zare, "Laser Techniques for Determinig State-to-State
     Reaction Rates" in State-to-State Chemistry, ACS Symposion
     Series 56, P.R. Brooks, E.F. Hayes Editors, Am. Chem. Soc.,
     Washington, D.C. 1977, p 50

13   E.J. Murphy, J.H. Brophy, G.S. Arnold, W.L. Dimpfl,
      J.L. Kinsey, J. Chem. Phys. 70, 5910 (1979)

14   D.A. Case, G.M. McClelland, D.R. Herschbach, Molec. Phys. 35,
      541 (1978)

15   D.S. Perry, A. Gupta, R.N. Zare, Electro-Optic Laser 80,
      Proceedings, Industrial and Scientific Conference Mangement
      Inc., Chicago (1981)

16   R.N. Zare, Polanyi Memorial Lecture, Faraday Discussion of
      the Chem. Soc. 67, 7 (1979)

17a   A. Gupta, D.S. Perry, R.N. Zare, J. Chem. Phys. 72, 6250 (1980)
  b   A. Gupta, D.S. Perry, R.N. Zare, J. Chem. Phys. 72, 6237 (1980)

18a   Z. Karny, R.N. Zare, J. Chem. Phys. 68, 3360 (1978)
  b   A. Torres-Filho, G. Pruett, J. chem. Phys. 77, 740, (1982)

19   R.W. Solarz, S.A. Johnson, R.K. Preston, Chem. Phys. Letters
      57, 514 (1978)

20   Z. Karny, R.C. Estler, R.N. Zare, J. Chem Phys. 69, 5199
      (1978)

THE EFFECT OF VIBRATIONAL AND TRANSLATIONAL EXCITATION

IN ATOM-MOLECULE REACTIONS

K. Kleinermanns and J. Wolfrum

Max Planck-Institut für Strömungsforschung
D-3400 Göttingen, Böttingerstraße 4-8
and
Physikalisch-Chemisches Institut der
Ruprecht-Karls-Universität Heidelberg
D-6900 Heidelberg, Im Neuenheimer Feld 253

ABSTRACT

An extensive review of experimental and theoretical investigations on reactions of vibrationally excited molecules is given in reference[1]. The paper here describes experimental investigations using UV lasers for selective translational excitation in gas phase radical reactions.

The nascent OH rotational, vibrational and fine structure state distributions produced in the endothermic reactions

$$H + O_2, \; CO_2, \; H_2O \longrightarrow OH \; (K,v,f) + O, \; CO, \; H_2$$

were measured by laser induced fluorescence at collision energies of around 251 kJ/mol with hot H atoms from the photodissociation of HBr at 193 nm. By monitoring reactant and product densities at short times, absolute cross sections of

$$\delta_{H-O_2} = 0.42 \pm 0.2 \; \text{Å}^2, \qquad \delta_{H-CO_2} = 0.37 \pm 0.1 \; \text{Å}^2,$$
$$\delta_{H-H_2O} = 0.24 \pm 0.1 \; \text{Å}^2$$

were obtained.

INTRODUCTION

The technique of flash photolysis for production of hot atoms in conjunction with <u>time-integrated</u> product detection methods has long been used to get information about reaction cross sections and excitation functions. Especially the reactions of hot hydrogen atoms and their isotopes with saturated hydrocarbons, D_2, CO_2, N_2O and others have been investigated. In all these experiments the measured reaction yields could only be deconvoluted to reaction probabilities or cross sections with guesses about the collisional cooling processes of the hot atoms. However, for most chemically interesting reactions, little is known about the detailed inelastic scattering processes.

EXPERIMENTS

We introduce a new method to study the dynamics of high barrier reactions combining hot atom and radical production by laser photolysis (high monoenergetic collision energies) and <u>time-resolved</u> product detection by laser induced fluorescence or other sensitive time- and state-resolved detection techniques. Here investigations on the endothermic reactions

$$H + O_2 \qquad OH (v,K,f) + O \qquad , \Delta H = 16.6 \text{ kcal/mol} \qquad (1)$$

$$H + CO_2 \longrightarrow OH (v,K,f) + CO \qquad , \Delta H = 25.4 \text{ kcal/mol} \qquad (2)$$

$$H + H_2O \longrightarrow OH (v,K,f) + H_2 \qquad , \Delta H = 14.8 \text{ kcal/mol} \qquad (3)$$

are described. These reactions are important in hydrocarbon combustion and in the CO oxidation as chain branching steps. We

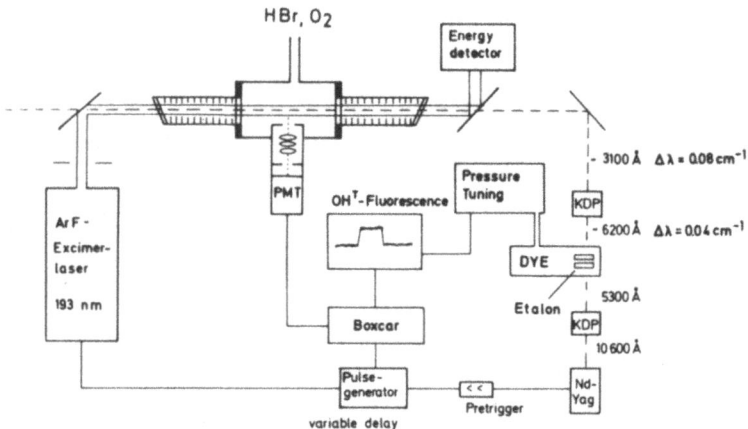

Fig. 1. Schematic of the experimental arrangement

report nascent OH rotational, vibrational and fine structure state distributions as well as absolute reaction cross sections for the three reactions.

The apparatus used is shown in Fig. 1. The OH state distribution is measured by combining Excimer laser photolysis of rapidly flowing HBr and O_2 mixtures with laser induced fluorescence detection of the OH product at a short time after the photolysis pulse scanning the $OH(X^2\Pi) \longrightarrow OH(A^2\Sigma)$ transition from 3060 to 3160 Å.

Normally relative peak heights above the base line were converted to relative OH densities by dividing the intensity by the laser power and the appropriate Hönl-London^2 and Franck-Condon factors^3. In some of the spectral scans the OH absorption was driven to saturation so that the LIF intensity is independent of variations of the laser power and directly proportional to OH density.

For the reaction $H + O_2$, only the v,K levels shown in Fig. 2 could be observed by LIF due to predissociation in the upper OH ($^2\Sigma$) state. The data for this reaction have been corrected for predissociation^4. The points K = 30 in v = 0 and K = 27 in v = 1 are necessarily zero by energy conservation for the $H-O_2$ reaction. The narrow dip in the v = 0 distribution does result from collision induced predissociation in OH ($^2\Sigma$) in the region of the $^4\Sigma^-$ curve crossing^4 and not from reaction.

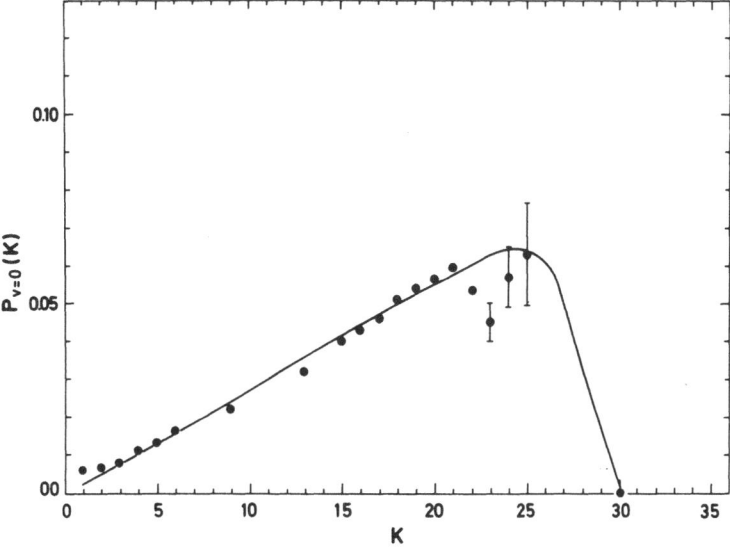

Fig. 2. Comparison of the measured rotational state distribution (♦) in OH (v = 0) produced in reaction (1) with the phase space statistical distribution (——).

A. Rotational Partitioning

At collision energies of 60 kcal/mol all three endothermic reactions are relatively fast. Typically OH spectra could be taken at a probe time of 100 ns at typical partial pressures 5 mtorr HBr, 50 mtorr O_2, CO_2, H_2O. At these pressures the gas kinetic collision frequency between the hot H atoms and O_2 is around 5×10^6 s^{-1}. At a probe time of 100 ns, less than one collision should have occurred between the hot hydrogen atoms and the gas and less than 0.1 collisions between the OH product molecules and the gas.

Fig. 3 presents typical portion of spectra obtained in the R-branch head of OH, v = 0 under approximately single collision conditions. The corresponding rotational state distributions are given in Fig. 4. Both probe times and total pressures were varied between 40-140 nsec and 50 and 200 mtorr without noticeable change in the form of the spectra due to rotational relaxation effects.

The most striking result is that the OH rotational distribution from the H-O_2 reaction is extremely hot, while the H-H_2O reaction leads to little rotational excitation despite comparable total energies of around 43 kcal/mol for both reactions. Obviously the dynamics and the potential surfaces for the two reactions are very different. The reaction $H(^2S) + O_2(^3\Sigma_g^-) \longrightarrow OH(^2\Pi) + O(^3P)$ is known to take place adiabatically on the ground state potential energy surface of $HO_2(^2A'')$. The HO_2 complex is 54 kcal/mol stable relative to the reactant region of the surface. Because of

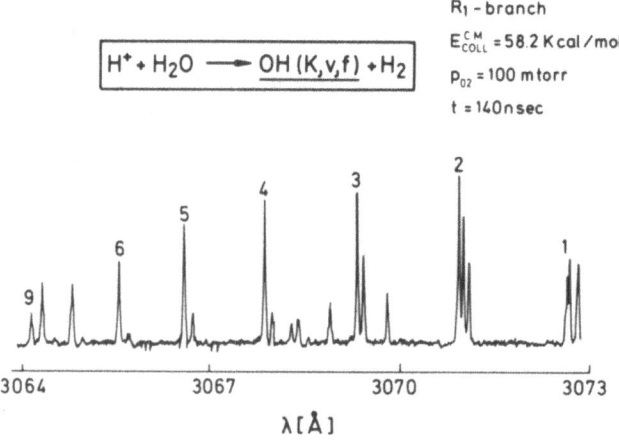

Fig. 3. Part of the rotational spectrum in OH (v = 0) produced in the reaction (3).

this deep HO_2 potential well it seems adequate to compare the measured rotational distribution with a statistical distribution. Since the reaction has angular momentum constraints, phase space theory[7] is the most appropriate statistical model for comparison with the experimental results. While the $H + H_2O$ reaction leads to a very nonstatistical rotational distribution, phase space and measured rotational distribution are in good agreement for the $H-O_2$ reaction (Fig. 2) over the measurable arrange of OH K-states ($K \lesssim 25$, $v = 0$).

However, care must be taken to conclude from this resemblance on the existence of a long living HO_2 complex at these high collision energies. We performed trajectory calculations on an ab initio $H-O_2$ potential surface[8] and got rotational distributions in good agreement with the experimental ones, but considerably hotter than the phasespace distribution in the high rotational energy range ($K < 27$, $v = 0$). The calculation showed us that no long living HO_2 complex exists at 60 kcal/mol collision energy. Typical residence[2] times in the complex region are on the order of 10^{-13} sec with only a few reflections (2-4 on the average) on the potential walls before entering product or reactant region respectively.

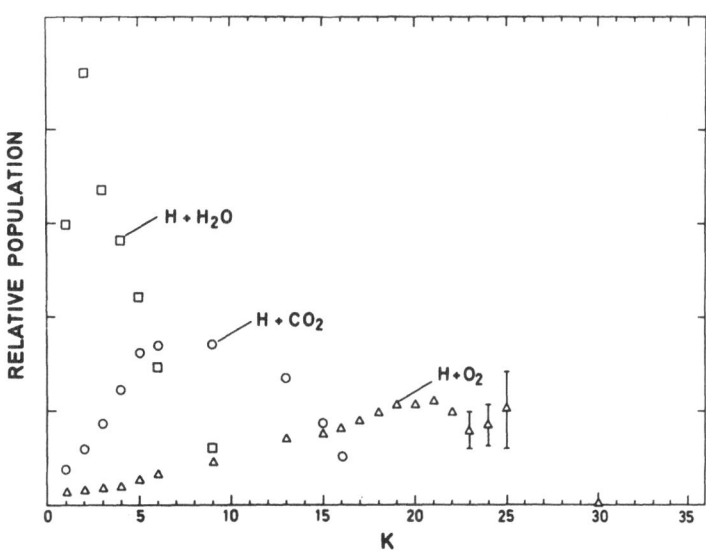

Fig. 4.   Nascent rotational state distributions in OH (v = 0) produced in the reactions (1)-(3). K is the rotational quantum number.

B. Vibrational Partitioning

Despite the high collision energies, no OH (v = 1) could be observed from the reactions H + H_2O and H + CO_2. Earlier investigations[11] have shown that the reaction OH + CO could not be vibrationally accelerated by excitation of OH to v = 1. Also for OH (v = 1) + $H_2 \longrightarrow H_2O$ + H the enhancement is less than a factor of 1.5[11] relative to the ground state reaction, while H_2 (v = 1) is reported[12] to react 150 times more rapidly with OH than H_2 (v = 0). It may be mentioned here that the OH bond distance remains almost constant going from OH to H_2O.

The reaction H + O_2 leads to substantial vibrational excitation. Since not all of the rotational states could be observed here due to predissociation, vibrational partitioning is taken from the observed linear rotational surprisals[10] in v = 0 and v = 1. The measured vibrational excitation $\delta(v = 1)/\delta(v = 0)$ = 0.45 $\pm$ 0.15 is somewhat lower than the phase space statistical excitation 0.69.

C. Fine Structure Partitioning

Spin-orbit and orbital-rotation interactions in the ^{2} state cause fine structure splittings for each rotational level. Each of these fine structure levels can be probed by different rotational subbands, i.e. Π_1^+ by P_1 or R_1, Π_1^- by Q_1, Π_2^+ by P_2 or R_2 and Π_2^- by Q_2, so that cross sections for production of the sublevels can be determined separately. The results are given in Table I. While the spin doublets are produced statistically, there

Table I   Observed fine structure partitioning for OH rotational levels produced in the reactions (1)-(3) at 60 kcal/mol collision energy.

	H + H_2O	H + CO_2	H + O_2
$\dfrac{\delta(^2\Pi_{3/2})}{\delta(^2\Pi_{1/2})} \cdot \dfrac{K}{K+1}$	1.1 $\pm$ 0.2	1.2 $\pm$ 0.2	1.2 $\pm$ 0.2
$\dfrac{\delta(\Pi^+)}{\delta(\Pi^-)}$	3.2 $\pm$ 1.0	3.0 $\pm$ 1.0	5.9 $\pm$ 1.0

is a preference for the $\Pi^{+}\Lambda$ -doublet component in all three reactions. The preference increases somewhat with K and approaches the values of Table I at the highest populated K.

For the rotational states, the electron density of the single occupied OH Π orbital forms lobes either in the rotation plane (Π^{+}) or perpendicular to it[14]. Adiabatically the Π orbital points along the direction of the bond that is broken during the course of the reaction. The results show that all three reactions generate a torque in a plane containing the bond to be broken, i.e. that the exit channels of the reactions occur in a plane. If the reactions could somewhat generate a torque about the bond to be broken, J_{OH} would be parallel to the Π orbital and the Π^{-} state preferentially populated. Preference for the OH Π^{+} state has been found before[15] in several chemical reactions leading to OH.

CONCLUSIONS

The combination of hot atom production and fast product detection seems to be a versatile tool to get information about the dynamics and cross sections of chemical reactions with substantial energy barriers. Nascent OH state distributions produced in the endothermic reactions of $H(^{2}S)$ with O_2, CO_2 and H_2O have been measured at 60 kcal/mol collision energy. Despite comparable total energies the OH rotational distributions from the reactions $H + O_2$ and $H + H_2O$ are drastically different implying different dynamics. The OH rotational distribution produced in the $H-O_2$ reaction is very similar to the statistical distribution, but quasi classical trajectory calculations show that the reaction proceeds very fast at 60 kcal/mol collision energy with only 0.1 psec residence time in the HO_2 complex region. In all three reactions the $\Pi^{+}\Lambda^{-}$ doublet component is preferentially populated demonstrating essentially planar exit channels.

Further experimental work will be directed to study the doppler profiles of the hot hydrogen atoms with a tunable L_{α} -laser light source[16] and to measure the reaction cross sections at different collision energies by varying the photolysis wavelength. The measurements will be compared with quasi classical trajectory calculations on ab initio and model surfaces to get more information about the widely unknown dynamics of high barrier chemical reactions.

ACKNOWLEDGEMENT

The financial support of the Deutsche Forschungsgemeinschaft is gratefully acknowledged.

REFERENCES

1. J. Wolfrum, "Reactions of Vibrationally Excited Mole-
    cules" Chapter 3 in "Reactions of Small Transient
    Species (A. Fontijn and the late M. A. A. Clyne,
    Editors), Academic Press, New York 1983
2. J. L. Chidsey and D. R. Crosley, J. Quant. Spectrosc.
    Radiat. Transfer, Vol. 23, pp. 187-199 (1980)
3. B. Lin, Private Communication
4. R. A. Sutherland and R. A. Anderson, J. of Chem. Phys.
    58, 1226 (1973)
5. R. S. Mulliken, J. Chem. Phys., 8, 382 (1940)
6. S. N. Foner, R. L. Hudson, J. Chem. Phys., 36, 2681
    (1962)
7. P. Pechukas, J. C. Light and C. Rankin, J. Chem. Phys.
    44, 794 (1966). Calculated assuming b_{MAX} = 1.8 Å ob-
    tained from trajectory calculation at 60 kcal/mol colli-
    sion energy on an ab initio HO_2 surface
8. C. Follelius and R. J. Blint, J. Chem. Phys. lett. 64,
    183 (1973)
9. G. C. Light, H. Matsumoto, J. Chem. Phys. lett. 58, 578
    (1979)
10. R. Bernstein and R. D. Levine, "Advances in Atomic and
    Molecular Physics", Vol. 11, p. 215, edited by
    D. R. Bates and B. Bederson, Academic Press, New York
    1975
11. J. E. Spencer, J. Endo, G. P. Glass, 16th Int. Symp.
    Combustion, p. 829, The Combustion Institute, Pittsburgh
    1977
12. R. Zellner, J. Phys. Chem. 83, 18 (1979)
13. H. M. Crosswhite and G. H. Dieke, J. Quant. Spectrosc.
    Radiat. Transfer 2. 97 (1962)
14. W. D. Gwinn, B. E. Turner, W. Miller Goss and
    G. L. Blackman, Astrophys. J. 179, 789 (1973)
    S. Green and R. N. Zare, Chem. Phys. 7, 62 (1975)
15. A. C. Luntz, J. Chem. Phys. 73, 1143 (1980)
16. R. Schmiedl, H. Dugan, W. Meier and K. H. Welge,
    Z. Phys. A 304, 137 (1982)

IR LASER-INDUCED DESORPTION,

REACTION, AND IONIZATION PROCESSES AT SURFACES

Peter Hess

Institut für Physikalische Chemie der
Ruprecht-Karls-Universität Heidelberg
D-6900 Heidelberg, Im Neuenheimer Feld 253

INTRODUCTION

A detailed understanding of the processes taking place at surfaces of condensed media is of fundamental importance in chemistry and physics. Often several processes contribute to a phenomenon and this complicates the analysis considerably. A catalytic reaction, for example, will proceed at length only if, in addition to the heterogeneous reaction, adsorption and desorption processes occur.

Experiments have been performed on pure substrates and on substrates covered with adsorbed layers of selected molecules. Two extreme solid substances, as far as the absorption of infrared radiation is concerned, are metals which possess very large absorption coefficients ($\sim 10^4 \text{cm}^{-1}$) and alkali halides which belong to the most transparent solids in the mid infrared ($\sim 10^{-3} \text{cm}^{-1}$). Of special interest at present are semiconductor materials with intermediate absorption coefficients ($\sim 10^{-1} \text{cm}^{-1}$).

A variety of methods is used to analyse surface processes. One of the most important techniques is mass spectrometry. Many other spectroscopic methods are also applied, such as surface enhanced Raman scattering or photoacoustics. The goal of such surface studies is the simultaneous application of several methods complementary to one another to elucidate the complicated surface processes.

DESORPTION

A point of considerable interest in surface science is if a
difference can be found between thermal desorption and laser
desorption of molecules induced by direct excitation of molecule-
surface vibrations with an infrared laser. Despite ultrafast
energy transfer to internal vibrational modes of the adsorbed
molecule and to surface phonons causing localized heating, such a
direct photon deposition may yield a high spectral selectivity and
a high efficiency. No experimental results are available for this
very interesting case.

Several papers have been published recently, however, where
the laser-stimulated desorption is studied by resonant excitation
of internal vibrational modes of adsorbed molecules using a CO_2
laser. For the molecular desorption of CH_3F from multilayer CH_3F
films condensed on NaCl (100) crystals at 77 K, a sharp maximum is
found around 990 cm^{-1}. The frequency dependence of the mass
spectrometer signal (CH_3F^+ ion current) nearly coincides with the
linear transmission spectrum of the excited v_3 mode obtained for
CH_3F adsorbed on NaCl. Despite the fact that about 3 CO_2 laser
quanta are needed to overcome the activation barrier for desorption, a relatively high quantum yield of 10^{-2} has been estimated[2].
This indicates a high efficiency of laser-induced desorption by
adsorbate vibrational excitation, elucidating the role played by
relaxation processes.

A selective frequency dependence of the desorption yield
was achieved not only with NaCl or KCl crystals, which are highly
transparent for CO_2 laser radiation and, therefore, minimize
substrate heating. A wavelength dependence has been observed, for
example, for the desorption of pyridine from KCl crystals[3], but
also from silver surfaces[4]. For pyridine molecules adsorbed on
silver surfaces, surface enhanced Raman scattering shows a band
centered around 1032 cm^{-1} with a bandwidth of about 8 cm^{-1}. The
major desorption occurs in the frequency range between 1025 cm^{-1}
and 1042^{-1}, indicating that direct excitation of the corresponding
v_8 asymmetric ring mode plays a major role[4]. However, not only
vibrational excitation of adsorbed molecules, but also absorption
of laser light by the silver film and the underlying SiO_2 is
responsible for desorption. The contribution of substrate heating
increases rapidly with laser intensity and allows laser desorption
at frequencies not resonantly absorbed by pyridine molecules.

The simple model of resonant absorption of the number of
photons needed to break the molecule-surface bond by an internal
vibration of the adsorbate and efficient transfer of this energy
to the surface bond is not a realistic description. Intramolecular
and intermolecular energy exchange in the adsorbate layers and
direct excitation of the substrate yields a region with a higher

degree of excitation at the surface. Therefore, not only originally excited molecules, but also neighbouring species are
desorbed and an isotopic selectivity cannot be expected. The
extent of this transient excitation is governed by the properties
of the condensed phase, such as heat capacity and dynamics of
energy exchange, and, of course, by the properties of the radiation source. The laser-induced desorption process takes place
during the short period of transient localized excitation before
the energy is dissipated into a larger region.

A frequency dependence of the desorption yield can be expected as far as the energy deposition is governed by wavelength dependent absorption of the adsorbate and not by substrate
heating, which often possesses a very small wavelength dependence.
The energy deposited in the condensed phase can be measured
directly by photoacoustic experiments. With this method the sound
waves are detected which are generated by the radiationless decay
of the laser excited states. In fact, just recently, a wavelength
dependence of energy deposition in the adsorbate has been demonstrated for submonolayer and multilayer surface coverages of SF_6
on silver films at 90 K by infrared laser photoacoustic spectroscopy[5]. The photoacoustic spectrum resembles the v_3 spectrum of
solid SF_6 at low temperatures and sharpens with increasing SF_6
coverage. The photoacoustic signal due to substrate absorption is
of course large for a metal substrate and, therefore, a stabilized
laser is needed for such experiments.

Several applications of laser stimulated desorption of
species from a surface have been described in the literature. Such
an application is cleaning of a surface with laser radiation.
Because very thin layers can be heated, desorption is favoured
against bulk diffusion and laser-induced desorption may be a
unique method for in situ cleaning of surfaces. However, the local
heating effect may induce chemical reactions and thereby change
the surface composition. This has been studied for CO adsorption
on Inconel 600, a Cr-Ni alloy of technological interest[6]. These
experiments were performed with a pulsed ruby laser and the sample
was heated up to 1700 K during irradiation. A rapidly growing
chromium oxide layer due to laser irradiation was found by monitoring the surface composition by Auger electron spectroscopy. At
a low partial pressure of CO in the experimental chamber the
oxide layer could be removed effectively by laser cleaning shots.

Soft laser desorption of molecules with a low vapour pressure
can be employed to increase the signal in a mass spectrometric
analysis. The drastic enhancement of the signal has been demonstrated by comparing a normal electron impact spectrum recorded at
10^{-8} Torr gas pressure with the enhanced spectrum obtained by
additional pulsed laser irradiation of a condensed sample of this
species during the scan through the mass range of interest[7]. At

higher laser intensities, ions can be produced at the surface and
a mass spectrometric analysis is possible without using the ionization source of the mass spectrometer. Normally, however, the
number of ions generated by laser irradiation is very small compared to the number of neutral species desorbed from the condensed
phase. In this case, ionization of laser desorbed neutral species
in the gas phase by chemical ionization, electron-impact ionization, or photoionization gives much higher ion yields.

REACTION

The influence of infrared laser radiation in promoting surface processes is not restricted to desorption of molecules. The
activation barriers of many surface reactions such as dissociative
chemisorption on metals are in the range of vibrational energies.
Therefore, it should be possible to enhance heterogeneous reactions by vibrational activation. A controlled excitation of
selected species can be achieved in the gas phase employing laser
irradiation parallel to the surface. In this configuration the vibrational excess energy may relax by collisional deactivation
before the excited molecules reach the surface. This will be the
case for a larger distance between laser beam and surface, and
efficient vibrational relaxation. No influence of the laser excitation on surface reactivity can be monitored under these conditions.

It has been shown that concentration changes in the gas phase
can be monitored via the intensity of the photoacoustic signal
generated by vibrational excitation. The kinetics of H-D exchange
between H_2S and D_2S, for example, can be detected by isotope specific excitation of D_2S or HDS using a CO_2 laser in parallel configuration and a photoacoustic concentration analysis[8]. The exchange reaction proceeds at the surface and it can be easily shown
that the kinetics of the heterogeneous reaction is not influenced
by the vibrational excitation used for photoacoustic analysis of
the process in the gas phase[9]. The effect of gas phase vibrational
excitation upon chemisorption of methane on thin films of rhodium[10]
has also been investigated employing the parallel configuration.
Excitation of the first vibrational state of the v_3 mode of CH_4
with a 3,39 μm He-Ne laser provides an energy of about 36 kJ/mol,
which is comparable to the activation energy of about 29 kJ/mol.
Nevertheless, the reaction probability estimated for CH_4 in the
first excited state of v_3 is below 10^{-4}.

An enhancement of the surface reactivity by vibrational excitation has been found in the gas-solid system SF_6-Si employing a
parallel and perpendicular configuration[11]. The interaction of
vibrationally excited SF_6^* or, at higher laser power, the reaction
of dissociation fragments of SF_6 with solid silicon yields vola-
tile SiF_4 which can be detected with a mass spectrometer. For per-

pendicular incidence of the laser beam on the solid surface the effect is more pronounced. However, in this configuration sub-strate excitation must also be considered. No reaction is observed in this configuration for CO_2 laser lines not absorbed effectively by SF_6. If, however, excited SF_6^* molecules are present, the additional excitation of lattice vibrations may increase the migration of adatoms and the desorption of product species. These effects may contribute to the enhancement of the surface reaction in the perpendicular configuration. In addition, heating of the silicon substrate seems to promote the surface reactivity.

Recently, experiments have been reported where laser-heating of the substrate is the main process responsible for the surface reaction. In this work the growing of SiO_2 oxide films by CO_2 laser-heating of silicon wafers in the presence of pure oxygen was investigated[12]. The advantage of this oxidation technique are the deposition of spatially localized layers and the excellent temperature control during the experiment. This allows forming very thin silicon oxide deposits on preheated silicon wafers, which can be used as rectifying diodes.

One of the most important branches of gas-surface chemistry is catalysis. Catalysis is a complicated, continuous process, where several steps are involved such as adsorption of species from the gas phase, reaction at the surface, and removal of the reaction products from the surface. It has been shown that an infrared laser can be used as an analytical tool to monitor cata-lytic processes without disturbing the heterogeneous reaction, or as a device which allows to influence individual steps. It is often difficult to connect an enhancement of the surface reac-tivity produced by laser irradiation to a distinct step in the series of catalytic processes. For example, vibrational excitation may promote the surface reaction; however, excited species also possess higher desorption rates than unexcited molecules. Thus, a better understanding of the laser-induced processes will be one of the important goals of future work in this field.

IONIZATION

A direct ionization of adsorbed polyatomic molecules by multi-photon absorption of infrared photons can be excluded in most cases. For multiphoton ionization of CH_3OH, for example, about 90 CO_2 laser quanta would be needed to reach the ionization con-tinuum. Dissociation of molecules into neutral species or radicals requires much smaller amounts of energy, e.g. about 30 CO_2 laser quanta to break the H_3C-OH bond. Therefore, ionic channels[2] should be of minor importance from an energetic point of view. In addi-tion, the pumping of very high lying states is extremely difficult in the condensed phase, because energy exchange processes are very efficient.

Nevertheless, the irradiation of sample layers of inorganic, organic, and bioorganic molecules with a CO_2 laser often leads to the formation of ions which can be detected with a mass spectrometer. This laser-induced ionization technique often allows the detachment of polar, thermally labile compounds of high mass from the surface without significant fragmentation. To avoid further fragmentation the ionization source of the mass spectrometer is normally turned off during these experiments. The dominant processes are alkali ion capture by polar organic molecules (the so called cationization)[13] and the formation of protonated or deprotonated molecular ions. The ionization efficiency decreases from polar to non-polar molecules and is normally in the range of $10^{-5} - 10^{-8}$[14].

Depending on the system under investigation and experimental conditions, different ionization mechanisms have been proposed. For example, ionic dissociation of organic salt molecules may occur under purely thermal conditions[15]. Alkali ions can be formed by thermal ionization of alkali contaminants present in nearly all types of samples. Cationization of polar molecules will then follow at the surface or in the gas phase. Proton transfer at H-bonds may lead to the preformation of ions in the condensed phase, which are detached from the surface together with neutral molecules. Short pulses of high power density can create a plasma containing atoms, molecules, fragments, and polymeric species in the neutral or ionized state. The ion spectrum detected under plasma conditions is determined by a series of secondary processes following the generation of the first ions. Altogether the mass spectra depend on many parameters such as laser wavelength, pulse duration, nature and preparation of the sample, ion-extraction geometry, etc.

From the many systems studied by laser desorption mass spectrometry of organic molecules only a few examples can be discussed here in more detail. Already in one of the first pioneering papers it has been shown that molecules with a very high molecular weight such as digitonin (MW 1228) can be analysed by laser desorption mass spectrometry[15]. The ionic species detected were the quasi-molecular ion $(M+Na)^+$ (m/e 1251) and the analogous lithium and potassium ions.

Laser-induced ionization has been observed not only with pulsed CO_2 lasers but also with cw CO_2 lasers. Two groups have shown recently that with a cw laser, areas of greatly different temperatures can be created at the sample or substrate. This can lead to independent thermal desorption of alkali ions from high temperature spots and evaporation of molecules such as sucrose from low temperature regions. The desorption spectra observed with the mass spectrometer are then predominantly due to ion-molecule gas phase reactions[16, 17].

Irradiation of a stainless steel target with a pulsed CO_2 laser yields not only singly charged, but also doubly charged[18] ions of the main constituents of the alloy. The detection of doubly charged ions and the similarity with a rf spark spectrum point to a plasma process. Condensation of small amounts of molecules such as D_2O, CH_3OH, or CH_3CN on the metal surface at 77 K decreases the threshold for ion detection considerably, so that in addition to the ion spectrum of the alloy, now fragment ions, molecular ions, and quasi-molecular ions of the condensed molecules are observed[18]. No dependence of the ion spectra on the CO_2 laser wavelength can be detected. Absorption of laser radiation by the metal substrate is the dominant process and resonant excitation of vibrational modes of the adsorbed species such as CH_3OH and CH_3CN is of minor importance. The influence of the metal sample[3] holder on laser-induced ionization disappears with increasing thickness of the adsorbed layer. No ions are detected for thick samples of CH_3CN even for efficient resonant vibrational excitation of the adsorbate. For thick methanol samples, however, infrared laser-induced ionization is observed at low laser fluences for resonant vibration excitation. The ion peak with the highest abundance belongs to the protonated monomer. Results obtained for the different methanol isotopes CH_3OH, CH_3OD, CD_3OH and CD_3OD reveal that attachment of the hydrogen isotope involved in the H-bonds is the main process[19]. These findings can be explained by laser-stimulated proton transfer at the H-bonds in the condensed phase and desorption of preformed ions[7]. Besides the main peak, some fragment ions and quasi-molecular ions such as the protonated dimer are detected and this shows that the processes induced by the laser are much more complicated. Sometimes a very clean laser desorption of preformed ions is possible. An interesting example is the mass spectrum of tetramethylammonium chloride, where the molecular ions $(CH_3)_4N^+$ and $[(CH_3)_4N]_2 Cl^+$ are detected, but no fragment ions[20].

An active area of current research is the application of laser desorption mass spectrometry (LDMS) to organic and biological substances[21]. Very often a striking similarity is found between the spectra observed by LDMS and those obtained by other mass spectrometric methods such as fast atom bombardment mass spectrometry (FABMS), secondary ion mass spectrometry (SIMS), and ^{252}Cf plasma desorption mass spetrometry (^{252}Cf-PDMS). If we accept that the central feature of all these methods is the production of a thermally activated region, irrespective whether photons or particles are used, and that the important ion formation processes are governed by the structural properties of the molecules (e.g. polarity, proton affinity etc.), these similarities can be expected. Characteristic differences in the ion spectra found for individual systems, however, can only be explained by a detailed understanding of the mechanism of ion formation and the reaction processes following this step.

REFERENCES

1. J. Heidberg, H. Stein, E. Riehl, and A. Nestmann, Z. Phys. Chem. N. F. 121 (1980) 145-164
2. J. Heidberg, H. Stein, and E. Riehl, Phys. Rev. Lett. 49 (1982) 666-669
3. T. J. Chuang, J. Chem. Phys. 76 (1982) 3828-3829
4. T. J. Chuang and H. Seki, Phys. Rev. Lett. 49 (1982) 382-386
5. F. Träger, H. Coufal, and T. J. Chuang, Phys. Rev. Lett. 49 (1982) 1720-1723
6. A. Pospieszczyk and J. A. Tagle, J. Nucl. Mater. 105 (1982) 14-22
7. M. Mashni and P. Hess, Appl. Phys. B 29 (1982) 205-211
8. R. Kadibelban and P. Hess, Appl. Optics 21 (1982) 61-64
9. P. Hess, R. Kadibelban, A. Karbach and J. Röper, to be published in the proceedings of the Third International Topical Meeting on Photoacoustics and Photothermal Spectroscopy, Paris (1983)
10. S. G. Brass, D. A. Reed, and G. Ehrlich, J. Chem. Phys. 70 (1979) 5244-5250
11. T. J. Chuang, J. Chem. Phys. 74 (1981) 1453-1460
12. I. W. Boyd and J. I. B. Wilson, Appl. Phys. Lett. 41 (1982) 162-164
13. F. Hillenkamp, Int. J. Mass Spectr, Ion Phys. 45 (1982) 305-313
14. P. G. Kistemaker, G. J. Q. van der Peyl, and J. Haverkamp, in Soft Ionization Biological Mass Spectrometry H. R. Morris ed., Heyden (1981) p. 120
15. M. A. Posthumus, P. G. Kistemaker, H. L. C. Meuzelaar, and M. C. Ten Noever de Brauw, Anal. Chem. 50 (1978) 985-991
16. R. Stoll and F. W. Röllgen, Z. Naturforsch. 37a (1982) 9-14
17. G. J. Q. van der Peyl, K. Isa, J. Haverkamp and P. G. Kistemaker, Org. Mass Spectr. 16 (1981) 416-420
18. B. Schäfer and P. Hess, Int. J. Mass Spectr. Ion Phys. 47 (1983) 47-50
19. M. Mashni and P. Hess, Chem. Phys. Lett. 77 (1981) 541-547
20. R. B. van Breemen, M. Snow, and R. J. Cotter, Int. J. Mass Spectr. Ion Phys. 49 (1983) 35-50
21. A. L. Burlingame, A. Dell, and D. H. Russel, Anal. Chem. 54 (1982) 363R-409R

STUDY OF MOLECULE-SURFACE INTERACTION BY LASERS

W. Krieger
Max-Planck-Institut für Quantenoptik
D-8046 Garching, Fed. Rep. of Germany

H. Walther
Max-Planck-Institut für Quantenoptik and
Sektion Physik, Universität München
D-8046 Garching, Fed. Rep. of Germany

SUMMARY

Combining surface-scattering experiments with laser-induced fluorescence allows the measurement of the internal energy distribution of surface-scattered molecules. As an example the experimental determination of the rotational state distribution of nitric oxide molecules scattered from a graphite and a Pt(111) surface is described and the results are discussed.

The detailed understanding of surface reactions is of large technical and scientific interest due to several reasons:

(a) Large-scale industrial processes are realized using catalytic reactions; important examples are ammonia synthesis, oil cracking, methanol production, catalytic combustion etc.
(b) In material science surface reactions are essential in connection with corrosion, steel hardening, or hydrogen embrittlement.
(c) The production of miniaturized electronic circuit elements on semiconductor surfaces makes extensive use of a variety of surface processes.
(d) Reactions between surfaces and electrolytes play a role in electrochemistry e.g. in fuel cells or in hydrogen production by photolysis.
(e) In biology and medicine surface reactions are of importance for e.g. bioadhesion or biomembranes.

A large number of methods has been developed to study surfaces. At the beginning the solid state properties of well defined surfaces and adsorbed layers were investigated and a
detailed picture of the atomic positions and electron structure
of many adsorption and coadsorption systems has been obtained.
This knowledge about the static system was the prerequisite for
the study of the dynamics of surface interactions and surface
reactions.

Information on the dynamics of a reaction may be obtained by
studying the reaction path, the dependence of the reaction cross
section on internal and translational energies of the reactants
and on the orientation of the reactants. Experimental investigations of reaction dynamics therefore require molecular beam type
experiments where no secondary processes as e.g. collisions influence the results. (This is not necessarily the case for studies
of reaction kinetics which uses rate constants to describe the reaction.)

Surface scattering experiments with atomic or molecular beams
also yield important information on the structure of surfaces and
of adsorbed overlayers, on surface phonons, surface potentials
and on the details of surface scattering. The subject has been
reviewed in a number of recent articles (see e.g. [1-4]).

The scattering process is governed by the surface potential
which generally consists of a short-range repulsive part and a
long-range attractive part forming an attractive potential well.
Several scattering mechanisms are possible:

(a) Elastic scattering. The angular distribution of elastically
    scattered particles usually consists of a very narrow scat-
    tering lobe in the specular direction.
(b) Inelastic scattering which leads to energy exchange with the
    surface (e.g. with surface phonons). The angular distribution
    of inelastically scattered particles is a much broader lobe
    not necessarily pointing into the specular direction.
(c) Trapping/desorption whereby particles are trapped in the sur-
    face potential well for a residence time τ and then desorb
    in a cosine distribution if energy accommodation on the surface
    was complete. The residence time may be calculated in a simple
    model as a function of the potential well depth E_d and of the
    surface temperature T_s:

$$\tau = \upsilon^{-1} \exp (E_d/k_B T_s) \qquad\qquad (1)$$

(d) Reactive scattering which includes processes like dissociative
    adsorption and catalytic reactions.

In many scattering processes a superposition of several of
these mechanisms is observed.

Scattering experiments with atoms or molecules are usually
carried out using effusive or supersonic nozzle beams. Well-
characterized and clean surfaces are necessary. Generally single-
crystal surfaces are used under ultrahigh vacuum conditions. The
scattered particles are detected with mass filters which also
allow time-of-flight measurements. In this way angular distributions, velocity distributions, residence times, sticking probabilities and reaction cross sections are obtained.

As long as only atoms are involved in the collision process the measurement of the angular distribution and the momentum change provides sufficient insight into the scattering dynamics. However, when molecules are scattered, additional information on the change of the internal energy is necessary.
Recently the laser-induced fluorescence method has been successfully used in several experiments to determine the influence of the surface interaction on the molecular rotational
distribution.[5-14] It has been shown that for the case of a
carbon-covered Pt(111) surface the rotational degree of freedom of the scattered NO molecules is only partly accommodated
to the surface temperature.[5,7] A similar effect has also been
observed for molecules which experience only weak inelastic
interaction with a solid surface, such as CO/LiF(001)[8] and
NO/Ag(111),[9,10] as well as for NO molecules desorbing from a
Ru(001) surface.[11] In the following, our own investigations of
the angular and rotational distributions of NO molecules scattered from different surfaces will be described in more detail.

In the experiments the laser-induced fluorescence of the
scattered molecules was measured in the $^2\Sigma \leftarrow {}^2\pi$ (0-0) transition
of NO before and after the scattering process. The electronic
ground state of NO is split into two states owing to spin-orbit
interaction, with the $^2\pi_{3/2}$ state about 120 cm^{-1} higher than
the $^2\pi_{1/2}$ state. The frequency-doubled radiation of an excimer-
pumped dye laser was used to excite the NO molecules (5 ns pulses
at a rate of 5 Hz, 10 μJ at 226 nm). The measured fluorescence
line intensities $I_{J'J''}$ allow the calculation of the relative population densities $N_{J''}$ of the ground state rotational levels according to:

$$I_{J'J''} \sim S_{J'J''} \, N_{J''} \, /(2J''+1) \qquad\qquad (2)$$

where $S_{J'J''}$ are the Hönl-London factors of the transition.

The setup for the laser-induced fluorescence measurement is
shown in Fig. 1. The UHV scattering chamber (10^{-10} torr) contained
a rotatable quadrupole mass filter for determining the angular

distribution of the scattered molecules. A supersonic NO beam with
a particle flux of about $2 \cdot 10^{14}$ molecules $cm^{-2}s^{-1}$ was scattered
from various surfaces. The number of $(NO)_x$ clusters produced during
the expansion in the nozzle is assumed to be negligible for the
chosen $p_o d$ product value of 3.5.[15] The details of the experiment
are described in Refs.[5,7,14]

For the molecules of the incident beam, the measured rotational populations show Boltzmann distributions in both fine-structure states with temperatures corresponding to 40 K ($^2\pi_{3/2}$) and
about 70 K ($^2\pi_{1/2}$). The second temperature also expresses the
ratio of the populations of the two electronic ground states. This
indicates that these states are not completely equilibrated with
each other by the collision processes during expansion in the
nozzle kept at room temperature. Assuming isenthalpic expansion,
one can estimate the translational energy of the incident molecules
from the measured rotational distribution to be about 700 cm^{-1} or
0.08 eV.

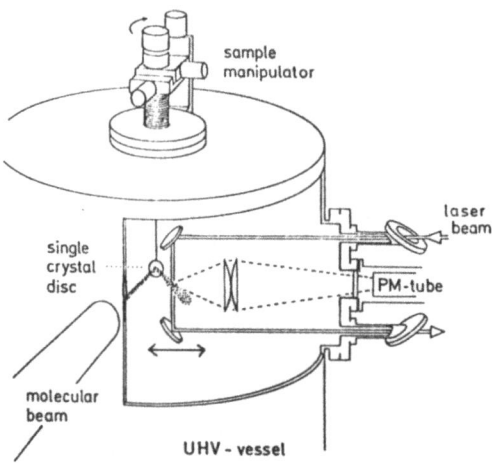

Fig. 1. Experimental arrangement (schematic). The laser beam
         enters and exits the vacuum chamber through Brewster
         angle quartz windows as shown. Inside the chamber it
         is deflected by aluminium coated mirrors which are
         displaceable to the left and right as shown, so that
         the incoming NO beam or the scattered molecules
         exiting the surface could be excited and observed.

At first the scattering experiments at a graphite surface will be discussed.[14] A polycrystalline graphite crystal of the Moore type was used whose properties were studied by He scattering.[16] According to these experiments a relatively high density of surface defects had to be assumed. For temperatures below 700 K chemical reactions between the surface atoms and impurity molecules in the beam (1 %, with main constituents N_2O, NO_2 and N_2) could be excluded. The upper limit of the NO coverage was estimated to be about 10^{-6} monolayers under steady-state conditions.

The rotational and angular distributions of the scattered NO molecules were investigated for surface temperatures between 130 K and 780 K. Examples of the measured angular distributions are shown in Fig. 2. Broad scattering lobes in the direction close to specular reflection, and underlying cosinelike distributions caused by diffusive scattering may be distinguished. The specular scattering lobe is interpreted as due to weakly inelastic scattering processes. The experimental results correspond qualitatively to the predictions of the hard cube model.[17,18]

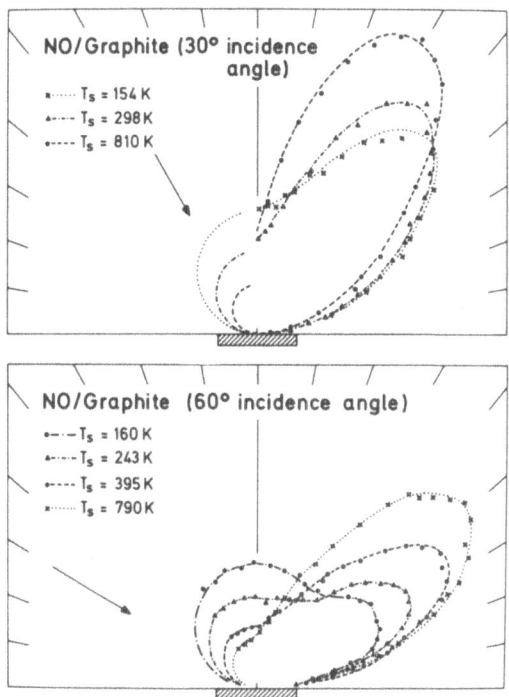

Fig. 2. Angular distribution of NO molecules scattered from a graphite surface at different temperatures with incidence angles of 30^0 (upper graph) and 60^0 (lower graph). The experimental points correspond directly to the mass-spectrometer signals.

As shown in Fig. 2 at increasing surface temperature the fraction of molecules scattered in a cosine distribution decreases while the specularly scattered fraction grows. A considerable fraction of the observed diffusively scattered particles, however, is thought to be due to the surface roughness of the crystal used. This is shown by the remaining isotropic part obtained at the highest temperature investigated.

The rotational distributions of the scattered molecules were measured at different surface temperatures for both electronic ground states. Within the range of rotational energies investigated and for all surface temperatures all rotational distributions could be fitted by Boltzmann distributions. They are therefore characterized by a rotational temperature. The same rotational temperature was found for both fine-structure states. The population ratio of the two states also corresponds to the same temperature. This means that the rotational levels and the two fine structure states are in thermal equilibrium after the scattering process.

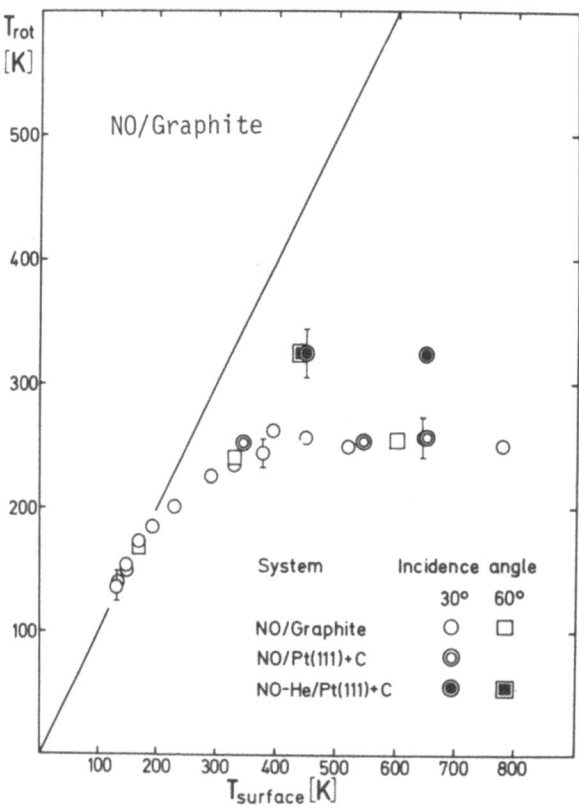

Fig. 3. Plot of the measured rotational temperature of the scattered NO molecules versus the surface temperature for the NO/graphite system. The straight line symbolizes full rotational accommodation.

The dependence of the rotational temperature on the surface
temperature is shown in Fig. 3. The solid line corresponds to
complete accommodation of the rotational degree of freedom to the
surface temperature. The experimental points follow this line up
to a surface temperature of 170 K, they deviate at higher temperatures and from 350 K upwards the rotational temperature converges
to a value of about 250 K.

The observation of specularly as well as diffusively scattered
NO molecules may be interpreted as a superposition of inelastic
scattering and trapping/desorption processes. As the surface temperature is raised the fraction of specularly scattered molecules
increases, indicating a predominance of inelastic scattering at
higher temperatures. A possible explanation for this behaviour is
the temperature dependence of the residence time τ of the NO molecules in the attractive potential well of the graphite surface.
With E_d = 0.12 eV[19] and υ = 10^{13} s^{-1} equation (1) yields residence
times between ~10^{-9} s for surface temperatures of 150 K, and ~1 ps
for T_s = 700 K. As the transit time of the molecules through the
potential well of the surface is estimated to be about 1 ps, the
molecules are assumed to undergo many hundred surface collisions
in the case of lower surface temperatures, leading to a higher
fraction of diffusively scattered particles. For higher temperatures the number of surface collisions decreases, resulting in a
growing specularly scattered fraction. The inelastic scattering
process responsible for this fraction leads to a redistribution
of the kinetic energy of the incoming molecules into the available
degrees of freedom of the system, i.e. kinetic and rotational
energy of the scattered molecules and surface phonons. The rotational energy of the scattered molecules is thus limited by the
amount of kinetic energy in the incoming molecular beam. This hypothesis was verified with experiments using a seeded NO beam whose
kinetic energy was increased by a factor of 2.5. As can be seen in
Fig. 3 this leads indeed to a higher rotational temperature.

For a second set of experiments[20] a Pt(111) surface was
used which was characterized with LEED. The potential well depth
of the system NO/Pt(111) is about 1.2 eV indicating a very strong
interaction of the NO molecules with this surface. As a consequence
below 300 K a chemisorbed layer of NO molecules corresponding to
about 0.3 monolayers is formed so that steady-state scattering
occurs mainly at the precursor potential of this adlayer. At about
330 K the majority of the NO molecules desorb and above 400 K the
NO coverage becomes negligibly small. The dissociation probability
of a NO molecule colliding with a Pt(111) surface is less than 10^{-3}.

Steady-state scattering experiments with NO from Pt(111)
resulted in angular distributions that closely follow cosine distributions for all surface temperatures investigated. The pre-
dominant scattering mechanism at the NO covered as well as the

clean surface is therefore trapping/desorption leading to full translational accommodation.

The measured rotational temperatures T_{rot} of the scattered molecules at different surface temperatures T_s are shown in Fig. 4. It is evident that full rotational accommodation is only obtained for $T_s < 300$ K, i.e. for the scattering at the precursor potential. At higher surface temperatures rotational temperatures much lower than T_s are found in spite of residence times of the order of ms and the observation of primarily diffusive scattering. Increasing the kinetic energy of the incoming NO molecules is without effect on the rotational temperature as is expected for a trapping/desorption process (see Fig. 4).

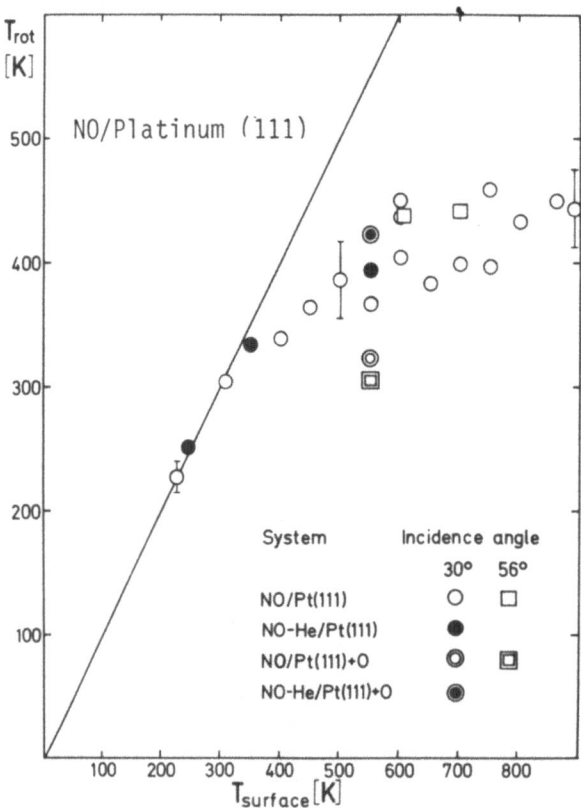

Fig. 4. Plot of the measured rotational temperature of the scattered NO molecules versus the surface temperature for the NO/Pt(111) system. The straight line symbolizes full rotational accommodation.

A similar deviation from full rotational accommodation has been obtained in a desorption experiment where a rotational temperature of 235 K was measured for NO molecules thermally desorbing from a 450 K Ru(001) surface, in spite of a very long residence time.[11] Assuming that there is no reactive interaction of the NO/Ru system, the result of the experiment shows that the final rotational distribution is predominantly determined by the exit channel of the interaction; the decisive processes leading to this final rotational distribution seem to take place when the molecules are receding from the surface. For the adsorbed molecules, rotation is hindered by the binding to the surface; thus, rotational energy is taken up when the molecules desorb from the surface.

The essential result we find in the present experiments is that within the range of rotational energies investigated the rotational distributions of NO molecules scattered from a graphite and a Pt(111) surface correspond to Boltzmann distributions of the same temperature for both electronic ground states. The measured angular and rotational distributions for the graphite surface are consistent with a superposition of two scattering mechanisms: trapping/desorption and inelastic scattering. Temperature-independent redistribution of the initial kinetic energy during inelastic scattering leads to a limit for the rotational temperature. Scattering from Pt(111) proceeds predominantly via trapping/desorption with full translational but only incomplete rotational accommodation. It is quite clear that the description of the process given here must be rough and incomplete and that a perfect understanding of the dynamics of the molecule-surface interaction should be obtainable by trajectory calculations.

The measurements show that laser-induced fluorescence is a powerful tool for the study of molecule-surface scattering. The experiments can easily be extended to study vibrational energy exchange at surfaces and to probe internal energy distributions of catalytic reaction products.

REFERENCES

1. J. P. Toennies, Appl. Phys. 3:91 (1974)
2. W. H. Weinberg in: "Advances in Colloid and Interface Science" Vol. 4, Elsevier, Amsterdam, 1975, p. 301
3. S. T. Ceyer, and G. A. Somorjai, Ann. Rev. Phys. Chem. 28:477 (1977)
4. B. Feuerbacher in: "Vibrational Spectroscopy of Adsorbates" R. F. Willis, ed., Springer, Berlin, 1980, p. 91

5. F. Frenkel, J. Häger, W. Krieger, H. Walther, C. T. Camp-
    bell, G. Ertl, H. Kuipers, and J. Segner, Phys. Rev.
    Lett. 46:152 (1981)
6. G. M. McClelland, G. D. Kubiak, H. G. Rennagel, and R. N.
    Zare, Phys. Rev. Lett. 46:831 (1981)
7. F. Frenkel, J. Häger, W. Krieger, H. Walther, C. T. Camp-
    bell, G. Ertl, H. Kuipers, and J. Segner in: "Laser
    Spectroscopy V", ed. by A. R. W. McKellar, T. Oka,
    B. P. Stoicheff, (Springer, Berlin, Heidelberg, New
    York 1981), p. 425
8. J. W. Hepburn, F. J. Northrup, G. L. Ogram, J. C. Polanyi,
    and J. M. Williamson, Chem. Phys. Lett. 85:127 (1982)
9. A. W. Kleyn, A. C. Luntz, and D. J. Auerbach, Phys. Rev.
    Lett. 47:1169 (1981)
10. A. C. Luntz, A. W. Kleyn, and D. J. Auerbach, J. Chem. Phys.
    76:737 (1982)
11. R. R. Cavanagh, and D. S. King, Phys. Rev. Lett. 47:1829 (1981)
12. L. D. Talley, W. A. Sanders, D. J. Bogan, and M. C. Lin, Chem.
    Phys. Lett. 78:500 (1981)
13. M. Asscher, W. L. Guthrie, T.-H. Lin, and G. A. Somorjai,
    Phys. Rev. Lett. 49:76 (1982)
14. F. Frenkel, J. Häger, W. Krieger, H. Walther, G. Ertl,
    J. Segner, W. Vielhaber, Chem. Phys. Lett. 90:225 (1982)
15. D. Golomb, R. E. Good, and R. F. Brown, J. Chem. Phys.
    52:1545 (1970)
16. D. L. Smith, and R. P. Merrill, J. Chem. Phys. 52:5861 (1970)
17. R. M. Logan, and R. E. Stickney, J. Chem. Phys. 44:195 (1966)
18. W. L. Nichols, and J. H. Weare, J. Chem. Phys. 63:379 (1975)
19. C. E. Brown, and D. G. Hall, J. Colloid Interface Sci.
    42:334 (1973)
20. F. Frenkel, J. Häger, W. Krieger, H. Walther, G. Ertl,
    H. Robota, J. Segner, and W. Vielhaber, to be
    published

LASER PHOTOLYSIS STUDIES OF QUINONE REDUCTION BY

CHLOROPHYLL A IN HOMOGENEOUS AND HETEROGENEOUS SYSTEMS

Francesco Castelli

Institute of General and Inorganic Chemistry
University of Rome
00100 Rome, Italy

ABSTRACT

Different systems containing chlorophyll a and quinones are
studied by pulsed laser photolysis as model systems for photosynthesis. The results obtained in fluid homogeneous solution are
discussed to elucidate the general mechanism of the photoreaction.
Heterogeneous systems, developed on the basis of this knowledge,
reproduce much more closely the properties of the "in vivo" systems.

INTRODUCTION

It is well known that photosynthesis is a process that converts
electromagnetic energy into chemical energy. Light is absorbed by
chlorophyll and in some special reaction centers (photosystems I
and II) the electronically excited states of chlorophyll produce
one electron transfers. The first products of the photoreaction are
therefore radicals which have a higher energy content than the
starting materials. These radicals produce through a series of dark
reactions the final products, O_2 evolution from water on one side
and CO_2 reduction to carbohydrates on the other side. As photochem-
ists, we are interested in the primary processes and therefore are
concerned with the formation and the decay of the species formed
following light absorption.

A possible and often convenient approach to the study of complex
biological systems is to investigate simpler systems as models with
the aim of obtaining some general mechanisms. These mechanisms may
relate to the functions of the more complex biological systems. The
systems, which are of interest to us and to several other laboratories, are chlorophyll-quinones systems. They are a good beginning

225

model for photosynthesis because the chlorophyll cation radical,
seen in natural photosystems, has also been detected in these systems.
      We have studied these systems by flash photolysis and followed
the kinetics of the transient species formed by following light absorption changes. On these experiments, samples are excited with a
brief intense pulse of light and change of light absorbance is monitored over time. Pulsed lasers are the ideal source for the brief,
intense exciting pulse needed in these experiments and lasers development has greatly improved the time resolution of the technique.
The machine used in our experiments was described by Castelli (1976).
It has several advantages because it used an N_2 laser pumped dye
laser rather than Neodymium or Ruby lasers. The good reproducibility
and high repetition rate make averaging possible and the lower energy
of the pulses greatly decreases the amount of damage to the sample.
This means that samples can be studied for long time and the experiments can be repeated at several wavelenghts. The dye laser also
makes possible to study the effect of varying the wavelenght of the
exciting pulse.
      The studies I am going to report on here were performed using
this laser photolysis apparatus in the laboratory of dr. Gordon
Tollin at the University of Arizona. They are part of a continuing
effort in many laboratories to develop models for photosynthesis.
I will first discuss the results in ethanol solution (the simplest
model system). These studies elucidate the general mechanism of the
chlorophyll photoreaction and can be used as a basis to study more
complex systems. These experiments lead to progressively more complex
systems and more complex models that are closer to the natural photo-
synthetic apparatus.

RESULTS AND DISCUSSION

Chlorophyll in Ethanol Solutions

      When a degassed ethanol chlorophyll a solution at room tempe-
rature is irradiated with visible light, a singlet excited state is
generated that has a lifetime of only few nanoseconds as measured
by fluorescence decay. However $\simeq$ 50% of the excited singlet decays
by intersystems crossing into a triplet state with a much longer
lifetime. Its decay can be easily detected by monitoring the change
of light absorbance at 465 nm as a function of the time from excitation, as shown in fig. 1 a). If benzoquinone is added at 10^{-4} -
10^{-3} M the triplet decay becomes much faster and a new slower
decaying species is observed (fig. 1b). The visible absorbance
spectrum of the new species (Chibisov, 1969; Kelly and Porter, 1970)
corresponds clearly to the π chlorophyll cation radical C^{+} produced
electrochemically (Borg et al., 1970).
      When the transient absorption spectrum is measured in the near
U.V. ($\leq$400 nm) a new species is detected with characteristic absor-

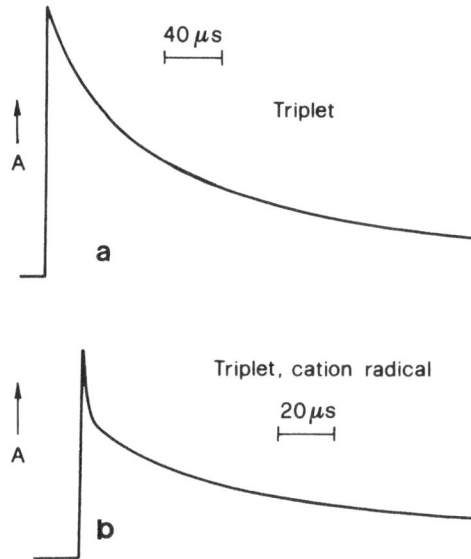

Figure 1 - Signal decay observed at 465 nm upon laser irradiation
           (655 nm) of degassed ethanol solutions: a) chlorophyll
           a alone; b) chlorophyll a + benzoquinone (10^{-4}M).

ption different from $C^{+}_{\cdot}$. The species should be the one that has
accepted the electron from chlorophyll and will therefore be a
quinone radical. In fact E.P.R. measurements at room temperature
show the characteristic signal from the semiquinone $Q^{-}_{\cdot}$ (Tollin and
Greens, 1962).

     The primary photochemistry therefore results in the production
of two radicals, a chlorophyll cation $C^{+}_{\cdot}$ and a semiquinone anion $Q^{-}_{\cdot}$.
The reverse reaction

$$C^{+}_{\cdot} + Q^{-}_{\cdot} \xrightarrow{k} C + Q$$

regenerating the original molecules, is inferred from the second
order decay of the slow signal in fig. 1b. We found the rate constant
for this process to be k=2x10^{9} M^{-1} s^{-1}, this agrees with the results
of Kelly and Porter (1970).

     From these absorption studies the mechanism, thus seem to be
straightforward. However further E.P.R. studies show things are not
so simple. The second order rate constant of the E.P.R. signal decay
was found to be 100 times smaller than k (Mukherje et al., 1969;
Hales and Bolton, 1972). Hales and Bolton (1972), have given a
good account of this problem and some of the proposed mechanisms.
Here I will discuss what we believe the mechanism is. All the results

are, in fact, easily re:cnuciled by including in the mechanism a
side reaction of C$^+$ with the solvent. If the rate constant for this
process is small deviations from the second order decay would be
difficult to see and only a small amount of excess Q$^-$ would be
produced. This species decays following second order kinetics according
to the disproportionation reaction.

$$2Q^- + 2H^+ \longrightarrow QH_2 + Q$$

and can only be detected by E.P.R.. We have obtained indirect evidence for the reaction of C$^+$ with the solvent from our work (Castelli et al., 1979) with pheophytin (chlorophyll with 2H$^+$ replacing Mg^{2+}).
In this case the pigment cation radical decays faster by mixed first
and second order kinetics leaving a large excess of Q$^-$ which can be
easily detected by laser photolysis. Since conspicuous isotope effects,
on the amount of excess Q$^-$, are observed when deuterated alcohols
are used as solvent, we can infer that the pigment radical ion reacts
directly with the alcoholic solvent. In order to gain some more insight into the chlorophyll photoreaction we have investigated how
temperature and the quinone redox potential affect the radical yield
and the reactions rate constants (Tollin et al., 1979). We found, as
expected, that both k_{qt}, the triplet quenching rate constant, and k
decrease with decreasing temperature because of the increase in
solvent viscosity. We also found the radical yield to decrease with
decreasing temperature and at -90 C it is ≃3 times less than at room
temperature. At this temperature, however, essentially all the triplet
still decays by the quenching process. This result clearly indicates
that free radical ions are not the direct product of the quenching
reaction and an ion radical pair must be formed first. This ion pair
can dissociate into separate ions or go back by reverse electron
transfer to the starting species. Decreasing temperature will favor
the reverse reaction. Changing a redox potential over a range of
0.85 V, by using various quinones, does not affect the second order
rate constant of the reverse electron transfer reaction. However
changing the redox potential does affect the rate of the triplet
quenching reaction and the radical yield. Even though the effect is
not very large it is quite clear from the results that the more oxi-
dizing the quinone is the faster the electron transfer rate is and
the higher the dissociation yield. Radical yield is also largely
affected by the solvent dielectric constant (Cheddar and Tollin, 1980).
    So far we have described experiments at quinone concentrations
(10^{-4} - 10^{-3}M) which would only affect the triplet state. If higher
concentrations of benzoquinone are used (10^{-3}- 10^{-2} M) singlet quenching can be observed by fluorescence measurements. In this range of
concentration, however, the total radical yield decreases. This is
a consequence of the fact that less triplet si formed because of
the competitive quenching at the singlet level. Quantitative data
support earlier conclusions that no free radicals are observed in
solution from singlet quenching in the chlorophyll-quinone system.
The solution system appears to be different from the "in vivo"

observation where there is evidence that the singlet is responsible
for the photochemistry. This matter has been reviewed by Parson and
Cogdell (1975).

At this point we can summarize in a scheme the reactions following light absorption in our simple model system.

$$C + h\nu \longrightarrow C_s$$

$$C_s \overset{k_f}{\nearrow} C$$

$$C_s \overset{k_{ic}}{\longrightarrow} C_t$$

$$C_s \overset{+Q}{\underset{k_{qs}}{\searrow}} (C^+ \ldots Q^-)_s \longrightarrow C + Q$$

$$C_t \overset{+Q}{\underset{k_{qt}}{\longrightarrow}} (C^+ \ldots Q^-)_t \overset{k_1}{\longrightarrow} C^+_\bullet + Q^-_\bullet$$

$$C_t \overset{k_{nr}}{\downarrow} C$$

$$(C^+ \ldots Q^-)_t \overset{k_1'}{\downarrow} C + Q$$

$$C^+_\bullet + Q^-_\bullet \overset{k_2}{\longrightarrow} C + Q$$

$$C^+_\bullet \text{ solvent } \overset{k_3}{\longrightarrow} C + \text{oxidized solvent}$$

$$2Q^-_\bullet + 2H^+ \overset{k_4}{\longrightarrow} Q + H_2Q$$

We can understand the striking difference in chemical behaviour
between the singlet and the triplet excited states by recalling that
the chemically formed radical pairs maintain the original differences
in spin character between the singlet and triplet. Since the ground
states, all have singlet character the reverse electron transfer
reaction is spin allowed for the singlet pair and diffusion is too
slow to produce free radicals. However, when the reverse electron
transfer reaction is spin forbidden as for the triplet radical pair,
diffusion becomes competitive and free radicals can be formed. Then
the radical yield depends on the relative rate constants k_1 and k_1'
of the two processes and is affected by the quinone potential and
the viscosity and polarity of the solvent.

There are two major differences between our first simple model
system and the "in vivo" system: 1) In nature the singlet excited
state is responsible for the photochemistry; in solution no products
can be observed from singlet quenching; 2) In nature the radicals
which are first formed after excitation generate final products
through a series of dark reactions; in solution the photo-radicals
go back to the starting materials with the exception of a small
amount of H_2Q. Thus one may ask: how does nature overcome the problem
of the back reaction of the radicals and the reverse electron

transfer in the singlet radical pair? Most people believe that the
answer must lie on the fact that the "in vivo" system has a space
ordered structure so that the electron goes through a series of
acceptors with gradually slightly higher oxidation potential so that
a large separation of the charges is obtained. If the first of this
acceptors is very close to the ion radical pair, then the electron
transfer to the acceptor may compete with and dominate the reverse
electron transfer even at the singlet level.

     If this explanation is correct and recent work in the picosecond
time domain (Fajer et al., 1980; Shuvalov et al., 1980) agrees with
this view, then it would be interesting to study the reaction in
viscous heterogeneous systems in order to slow diffusion and therefore the radical back reaction. In these systems we may be able to
separate different radicals in different phases. With this aims we
have studied systems with chlorophyll molecules incorporated in films
(Cheddar et al., 1980) and in liposomes (Hurley et al., 1980; Hurley
et al., 1981).

Chlorophyll in films of cellulose acetate

     When degassed film of cellulose acetate containing chlorophyll
are excited with the laser at 655 nm only one slow transient is
observed. This is easily identified as the triplet state. There are,
however, some interesting differences from solution which are a consequence of the different environment the chlorophyll molecules
experience. Linschitz and Sarkanen (1958) found that the chlorophyll
triplet in fluid solutions does not follow single exponential kinetics. They detected a second order component which they attributed
to a process of triplet-triplet annihilation. Furthermore the triplet
state was also quenched by chlorophyll ground states. The chlorophyll
triplet decay in the film is instead rigorously exponential with a
lifetime of 1.4 ms which does not depend on the concentration of
either the triplet or the ground state. If the film is soaked in
water, the lifetime becomes slightly faster. In both cases, dry
films and soaked films, the triplet decay is strongly quenched by O_2
just like in solution. Evidently, in the cellulose acetate film,
chlorophyll molecules are isolated from each other by being trapped
in the cages formed by the criss-crossing of the long polymer chains
of cellulose. At high pigment concentration (over 10^{-3} M) some cages
may contain more than one chlorophyll molecule. These small aggregates, even though they are in such a small concentration that they
cannot be detected by UV-Vis absorption spectra, are responsible
for the decrease in triplet yield, observed at high chlorophyll concentration. Energy is not directly absorbed by the aggregates in any
significant way. However it may be channeled to these particular
centers, by fast inductive resonance energy transfer. In other words,
most chlorophyll molecules are isolated so that triplet levels may
not collide with other chlorophyll molecules; however they are close
enough, so that singlet resonance energy transfer among them may be

fast enough to compete with intersystem crossing. If aggregates are present they work as energy traps. Energy migrates among chlorophyll molecules until a trap is reached; here quenching occurs lowering the triplet yield. The process we have just described is quite similar to the "in vivo" process where the electromagnetic energy, harvested by single chlorophyll molecules at high concentration, is transferred to the reaction centers. It should be noted that this process could not be observed in fluid homogeneous solution since, at such high chlorophyll concentrations, there is considerable aggregation.

The fact that the chlorophyll triplet in the film is easily quenched by O_2 shows that the film is porous and the chlorophyll cages can be reached through small channels. Therefore it was of interest to see if quinones could quench the chlorophyll triplet in the film and eventually produce radicals. In experiments where degassed 8 mM benzoquinone water solutions were added to chlorophyll containing film, only triplet quenching was observed when the laser photolysis measurements was made immediately after the addition. However, when measurements were made a few minutes after addition, then a slower component could be detected in the absorbance signal decay. The relative amplitude of the slow to the fast signal increased with the time after addition but a constant signal was observed after 20 minutes. If a water pre-soaked film was used, then this same signal was observed immediately.

The slow transient can be identified as due to the radicals by the absorption spectrum, by the fact that it decays by second order kinetics, and because it is not quenched by O_2. Furthermore an E.P.R. signal due to $Q^{\cdot}$ can be detected from these systems under red-light steady state irradiation.

The rate constant for triplet quenching is 3.7×10^6 $M^{-1}s^{-1}$ which is about 10^3 times smaller than in alcoholic solution. The rate constant for the second order back reaction is also decreased by the same factor. The absolutue radical yield from the triplet is decreased $\simeq 3$ times in the film with respect to the solution. All these differences from solution can be interpreted to be caused by the film structure because the channels to the chlorophyll cages are a high viscosity medium for the quinone molecules. The diffusion of these species through the maze of channels of an unswollen film is still fast enough to compete with the slow triplet decay rate, but once the radical pair is formed, $Q^{\cdot}$ can not diffuse away and reverse electron transfer takes place. When films are soaked, microviscosity is probably decreased by the water penetration and solvation of the interior part. In fact the film tickness increases several times with soaking. Chlorophyll cages may also become larger and the local dielectric constant may increase. Under these conditions a fraction of radical pairs may dissociate to yield free radicals. Then the $Q^{\cdot}$ radicals will diffuse through the maze of channels and cages until they find C^{+} radicals and the reverse electron transfer may take place. Thus the system can be described as a medium with high average viscosity through which the quinone species may diffuse, while

the chlorophyll species stay in fixed positions. Thus the electron
transfer reactions in the film and in the solution are similar in
both, their components and reaction order. In the film, however,
because of the high viscosity, rate constant are much smaller and
this may be useful for storing the chemical energy of the radicals.

Chlorophyll in liposomes

A somewhat different behaviour is observed when chlorophyll is
·dissolved in liposomes. These systems are commonly used as models
for biological membranes. They consist of spherical double layers
of lipids molecules which are arranged in space so that the polar
end is directed toward the inner solvent sphere or toward the
external solvent (water) in which the spherules are suspended. The
distribution of chlorophyll molecules among liposomes is of statistical nature. Therefore in the same sample the number of chlorophyll
molecules per liposomes N is not constant and only an average $\bar{N}$ can
be calculated by the relative amounts of chlorophyll and lipid and
the liposomes average size. For very low $\bar{N}$ (≤ 0.2) the fraction of
liposomes having more than one chlorophyll molecules is insignif-
icant and most liposomes have only one or no chlrophyll molecules.
Under these conditions chlorophyll molecules are isolated from each
other and, as it was for chlorophyll in the cellulose acetate films,
the triplet decay is exponential. However, at higher $\bar{N}$, when the
fraction of liposomes containing more than one chlorophyll molecule
becomes significant, the triplet decay becomes faster and not exponential. Since triplet quenching is a collisional process (see our
results with the film), chlorophyll molecules must be able to move
within the liposome so that ground state molecules can quench the
triplet excited states. The fact that the triplet decay is not
single exponential, but does not have second order components (no
changes are observed with changing laser intensity), is a consequence of the intrisic statistical heterogeneity of the sample. If
average decay rates are considered, a quenching rate constant of
2×10^5 $M^{-1}s^{-1}$ can be calculated. At high $\bar{N}$, as it was also observed by
Beddard et al. (1976) quenching takes place at singlet level by the
mechanism of inductive resonance energy transient to statistical
traps. Therefore, as in the films, triplet yield decreases.

If quinones are added, the chlorophyll triplet state is quenched
at a rate which, like in the films, is much smaller ($\simeq 10^2$ times)
than in alcoholic solutions. Apart from the value of the rate
constants, the processes observed in liposomes at the chlorophyll
triplet level are the same as the ones observed in solution or in
the cellulose acetate films, but important differences are found
in the radicals formation and decay.

If benzoquinone is used to quench the triplet, a composite
transient signal is observed which can be separated in two parts,
a first fast component with 150 μs halftime and a much longer one
with a halftime as long as 40 ms. The spectral properties of these

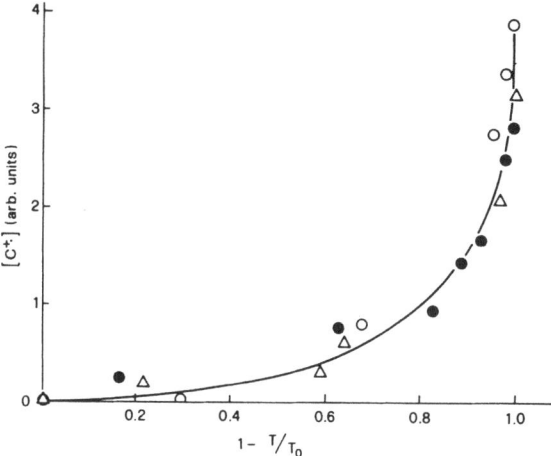

Figure 2 - Radical yield (C^+) in liposome samples containing
chlorophyll and quinones as a function of the fraction
of chlorophyll triplet quenched $(1 - T/T_0)$, where T and
T_0 are the triplet lifetimes of samples with and without
quinones. Quinones, used, are: benzoquinone (•) ubiquinone (o) and 2,6-d-t-butyl-p-benzoquinone (Δ).

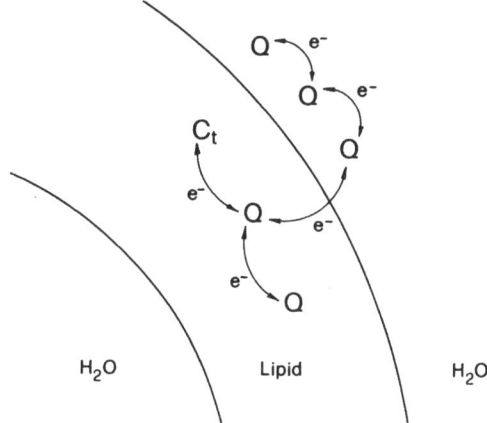

Figure 3 - Scheme for electron transfer chlorophyll-quinone and
quinone-quinone in liposomes.

transients correspond to those of the radicals. Both decays have
first order character since they are not affected by laser intensity.
If ubiquinone is used only the first fast component is observed.
Since ubiquinone, unlike benzoquinone, is insoluble in water, we
can conclude that the fast decay must be due to radicals recombination inside the liposome, and the long component must be due to the

recombination of C^+ inside and Q^- at the liposome-water interface.
This implies that electron transfer to the quinone molecules outside
the membrane must occur. Also, since radicals decays follow first
order kinetics, the semiquinone does not redistribute between liposomes. The electron transfer between quinone molecules is evidentia-
ted by the non linear dependence of the radical yield on the fraction
of triplet quenched. Indeed a large positive cooperative effect is
found (fig.2). The same dependence is found for both the amount of
C^+ which decays slowly in the cases of benzoquinone and 2,6-d-t-butyl-
-p-benzoquinone or the total $C^\ddagger$ which decays fast in the case of
ubiquinone. This means that the same mechanism of electron hoping
from quinone to quinone is responsible for moving the electron away
from the first radical pair and for moving the electron from inside
to outside the membrane. A scheme for the process is given in fig.3.
Here the reverse process is also shown: increasing Q concentration
will create more pathways for the electron to go back inside and
reach $C^\ddagger$. The rate of the slow process would increase and this is
in fact observed. On the contrary the rate of the fast decay is not
affected by the quinone concentration; evidently in the membrane
reverse electron transfer from Q^- to $C^\ddagger$ is the rate limiting step
of the all process. More evidence for the $Q \rightarrow Q$ electron transfer
process is gained by studying samples with two quinones. More pre-
cisely a series of samples containing ubiquinone and benzoquinone
were prepared. The concentrations were such that benzoquinone could no
compete with ubiquinone to quench the chlorophyll triplet. However,
increasing the benzoquinone concentration, the amount of slow
radical, which is produced only by the benzoquinone, increases.
Under the conditions used, no slow radical should have been produced
if no $Q \rightarrow Q$ electron transfer occurred. It should be noted that in
these last experiments the model system reproduced closely the "in
vivo" system. In fact the electron is moved away from the first
acceptor in the radical pair by quinone to quinone transfers to
reach a quinone which is outside the membrane and form a much more
stable radical. Obviously there are some important differences. The
electron transfer does not go through a space ordered series of
different acceptors, but it is rather random and only two kinds of
acceptors are used. Radical formations occurs again only at the
triplet level: in fact singlet quenching results in a decrease of
total radical yields. However, in comparison to the first model
system, the homogeneous fluid solution, it should be recognized
that the liposome system yields a much better model for the "in vivo" system.
    In conclusion, understanding the simplest model system, has
helped in producing better models and better systems. Molecular
compartimentalization of chlorophyll in viscous   and heterogeneous
media has allowed to generate systems where chlorophyll triplets
have little or no interactions with the other pigment molecules,
despite the large local chlorophyll concentration. Quinone-quinone
electron transfer in the membrane produces high radical yield, which
may be almost as large as in fluid solution, despite the much slower

rate of diffusion. The decay rate of the radicals has been decreased drastically and radicals have been separated in different phases.

     We are now in a position in which we can start to study ways to store some of the absorbed electromagnetic energy, using the radicals to produce, for instance, more stable chemicals. This will open a new chapter in photosynthesis modelling and may also produce some practical photochemical devices.

AKNOWLEDGMENTS

     Thanks are due to drs. Gordon Tollin, Frank Rizzuto, John K. Hurley and Glen Cheddar for all the fun we have had working at these problems.

REFERENCES

Beddard, G. S., Carlin, S. E. and Porter G., 1976, Chem.Phys.Lett.,
        43:27.
Borg, D. C., Fajer J., Felton R. H. and Dolphin D., 1970,
        Proc.Natl.Acad.Sci., U.S., 67:813.
Castelli, F., 1976, Chem.Phys.Lett., 38:528.
Castelli, F., Cheddar G., Tollin, G. and Rizzuto F., 1979,
        Photochem.Photobiol., 29:153.
Cheddar, G. and Tollin, G., 1980, Photobiochem.Photobiophys., 1:235.
Chibisov, A. K., 1969, Photochem.Photobiol., 10:331.
Fajer, J., Davis, M.S., Forman, A., Klimov, V. V., Dolan, E. and
        Ke, B., 1980, J.Am.Chem.Soc., 102:7143.
Hurley, J. K., Castelli, F., and Tollin, G., 1980, Photochem.
        Photobiol., 32:79.
Hurley, J. K., Castelli, F. and Tollin, G., 1981, Photochem.
        Photobiol., 34:623.
Kelly, J. M. and Porter, G., 1970, Proc.Roy.Soc.Sez.A, 319:319.
Linschitz, H. and Sarkanen, S., 1958, J.Am.Chem.Soc., 80:4826.
Mukherjee, D. C., Cho, D. H. and Tollin, G., 1969, Photochem.
        Photobiol., 9:273.
Parson, W. W. and Cogdell, R. J., 1975, Biochim.Biophys.Acta,
        416:105.
Shuvalov, V. A., Klimov, V. V., Dolan, E., Parson, W. W. and Ke, B.,
        1980, FEBS Lett., 118:279.
Tollin, G. and Green, G., 1962, Biochim.Biophys.Acta, 60:524.
Tollin, G., Castelli, F., Cheddar, G. and Rizzuto, F., 1979,
        Photochem.Photobiol., 29:147.

PART IV

APPROACHES TO LASER SYNTHESIS

LASER INITIATED FREE RADICAL CHEMISTRY

D. J. Perettie, J. C. Stevens and J. B. Clark

The Dow Chemical Company
Central Research
Midland, MI 48640

The topic which will be covered in the series will be that of the use of lasers in free radical processes. The object of our work is to either develop novel processes based on laser initiated free-radical reactions or optimize the present reactions. Many industrial chemicals are currently produced via organic photochemistry including commodity chemicals such as caprolactam and methylchloroform. There also have been recent reports[1] of a potential photochemical pathway for the production of vinylchloride which is another large volume material.

There are very few basic differences in the free-radical context between a laser and a standard mercury arc lamp. The major uniqueness of using a laser to study these reactions is power density; which can be important in some cases, geometrical considerations; which can become important when an engineer considers the problems with the transport of light through a long tube of small diameter, and finally, the ability to vary the wavelength; in an effort to study this variation on the chemistry in question.

It is first important to understand the basic type of chemistry being considered. The typical free-radical reaction consists of many separate steps, which are outlined below for a chlorination reaction.

$$R-Cl + h\nu \longrightarrow R\cdot + Cl\cdot \qquad \text{Initiation} \qquad \text{I}$$

$$Cl\cdot + R-Cl \longrightarrow \cdot RCl + HCl \qquad \text{Propagation} \qquad \text{II}$$

$$\cdot R-Cl \xrightarrow{\;Cl_2\ \text{addition}\;} RCl_2 + Cl\cdot \qquad \begin{array}{c}\text{(determined by}\\ \text{temperature)}\end{array} \qquad \text{III}$$

β-elimination
----------------->   R= + Cl·   IV

·R-Cl + Cl· -----> RCl$_2$ Termination   V

It is also possible to initiate via homolytic bond clevage of X$_2$(Cl$_2$).

Cl$_2$ + hν -------> 2Cl·   VI

The chlorine atoms, once formed, propagate via II as shown above.

The following discussion will entail two types of free-radical processes which proceed as above at temperatures such that both the addition and elimination products are observed.

1. The Chlorination of 1,1-Dichloroethane

This reaction proceeds by the addition route involving photoinitiation of chlorine with subsequent hydrogen abstraction and addition on the 1,1-dichloroethane. The overall reaction is:

$$CH_3CHCl_2 + Cl_2 \xrightarrow{h\nu} CH_3CCl_3 + CH_2ClCHCl_2 + HCl$$
                        VIIa   VIIb

As can be seen from equation VII, two major products result from this reaction. The advantage of photochemistry over standard thermal initiation is that the ratio of α-chlorinated product (VIIa) to β-chlorinated product (VIIb) is enhanced at lower temperature, i.e. ∿2.5/1 thermal and >4.0/1 photochemical (350 nm Hg) for the α/β chlorination ratio. It was also discovered by Mintz[2] that the addition of small amounts of iodine appeared to enhance the above ratio. Further studies on this observation indicated that the α/β ratio is not only additive dependent but also wavelength dependent as can be seen in figure 1. Upon further consideration of the mechanism it seems logical that an interhalogen is formed as an integral part of the propagation reaction, i.e.

Cl$_2$ + X$_2$ <=====> 2 X Cl   VIII

where the wavelength dependence closely follows the absorption maxima of the mixed halogen, XCl, as can be seen in figure 2.

Although such a process will never be initiated by a laser it seems obvious that having the high power variable wavelength sources available certainly facilitates such a study.

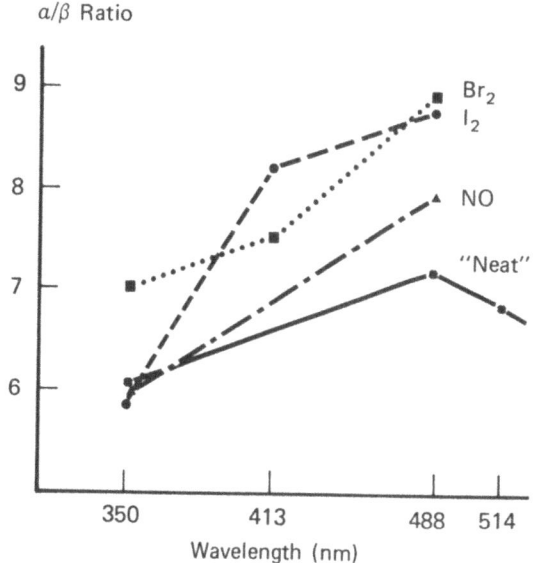

Figure 1. The wavelength and additive dependence on the chlorination at 1,1-dichloroethane.

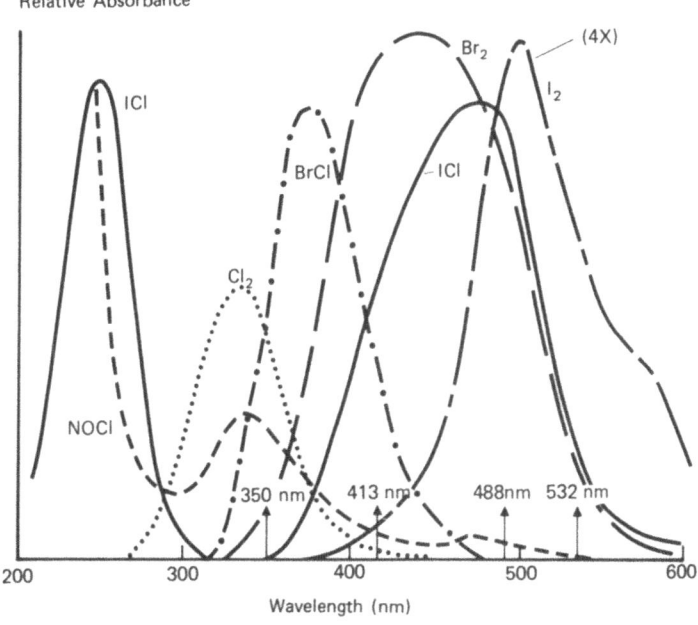

Figure 2. The wavelength dependence for the chlorination of 1,1-
dichloroethane.

2. The Dehydrochlorination of 1,2-Dichloroethane

This reaction proceeds via the β-elimination of Cl· after
H abstraction with Cl·. The reaction has been extensively
studied by Wolfrum[1], et.al., employing a KrF laser at 249 nm to
cleave the C-Cl bond in the chlorinated ethane;

$$CH_2Cl-CH_2Cl + h\nu \longrightarrow CH_2Cl-CH_2\cdot + Cl\cdot \qquad\qquad IX$$

the reaction then propagating via the elimination pathway
resulting in the production of vinylchloride.

The present work entails the initiation of the reaction
using Cl_2 and irradiation at about 350 nm. The resulting
propagation followed the pathway described above. Both pulsed
and c.w. sources were employed resulting in ∿50% conversion at
the powers studied. The table below summarizes the results.

Laser	Power	λ (nm)	Conv. (%)
Kr^+	1.80	350	49
Kr^+	1.40	350	49
Kr^+	0.80	350	45
Kr^+	0.02	350	49
N_2	0.03	337	50

As can be seen the conversion of 1,2-dichloroethane to
vinylchloride is quite independent of power in the range
studied. Obviously there are sufficient photons to initiate
the Cl_2. In addition to varying the laser power, the effect
of chlorine initiator concentration was studied as is shown in
figure 3. As can be seen from the plot, the variation of
pulsed vs. continuous source is nil. In addition, above
1000 ppm the increase in initiator has little effect.

This reaction also can be initiated with a conventional
source but it is felt that for geometric considerations it
would be advantageous to employ a laser. Based on the present
work, the authors would choose a N_2 or XeF laser of very low
power.

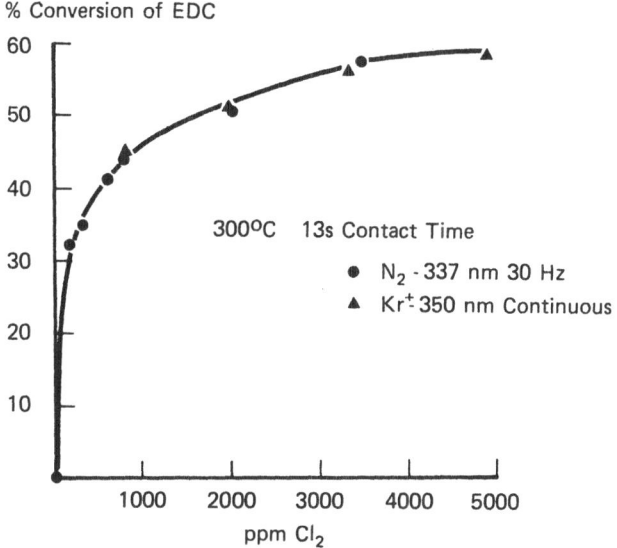

Figure 3.   The initial dependence of 1,2-dichloroethane decompo-
            sition.

References:

1.   J. Wolfrum, M. Kneba, P. Clough, German Patent P2938 353.7 (1979).

2.   M. Mintz, U.S.P. 3580831 (1971).

LASER INDUCED SELECTIVE DECOMPOSITION REACTIONS

D. J. Perettie, S. M. Khan, J. B. Clark and
J. M. Grzybowski
The Dow Chemical Company
Central Research
Midland, MI 48640

The work which will be reported in the area of selected decomposition encompasses many areas, each of which will be treated separately. The photon sources for the work which will be described entailed the use of lasers, but in many cases metal arc lamps undoubtedly could be employed.

A. The Photochemical Formation of CF_2O by CF_2HCl Decomposition

One of the areas which has been studied is the multiphoton infrared decomposition of CF_2HCl to generate CF_2, which subsequently reacts with O_2, ultimately producing CF_2O. The pressure dependence of both reactants was investigated and is displayed in Figure 1. The products were determined by GC/MS and FTIR.

The high yields experienced for CF_2O tend to indicate that the CF_2 carbene is formed in the triplet state resulting in an allowed reaction with O_2. Additional studies of the fluorescence during irradiation with a multichannel analyzer show an emission at 589.6 nm which has been assigned to the $^3B_1 \rightarrow {}^1A_1$.[1,2] This is an excellent example of a system which populates by inverse electronic excitation by multiphoton irradiation. The surprising fact to note is that product yields to 86% can be obtained by this technique.

B. The Selective Decomposition of 2,4,5-Trichlorophenoxyacetic Acid

Another area where selective decomposition reactions could play an important role is in purifications. It is possible to invision an impurity with a UV absorption spectrum somewhat different than the host material. A good example of this is depicted in Figure 2.

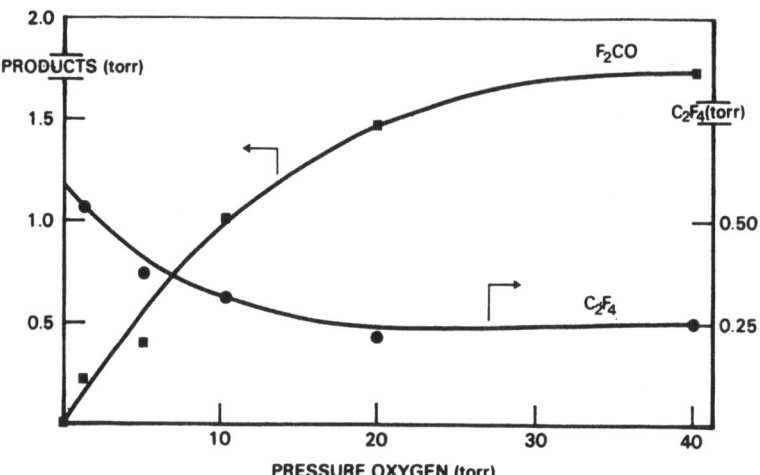

Figure 1.   Oxygen pressure of CF$_2$O formation.

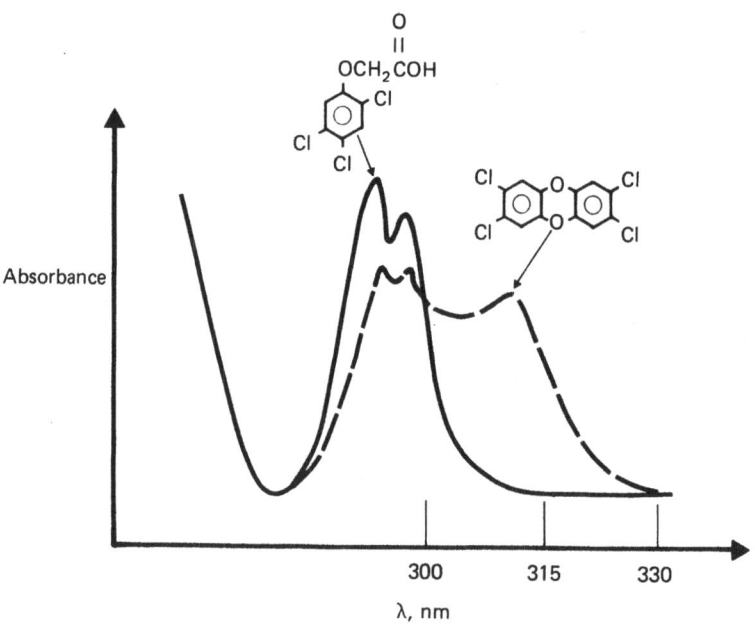

Figure 2.   The absorption spectra of Dioxin in 2,4,5-T

The difference in absorption between 308 nm and 330 nm indicates that the TCDD could possibly be decomposed in the presence of 2,4,5-T. This concept has been previously published[3] but no wavelength dependent data was reported.

A summary of the data gathered in the present study is contained below in Table I. The experiments at 308, 320 and 337 nm indicate that the optimum wavelength to selectivity decomposes TCDD is around 320 nm where the 2,4,5-T absorbs only slightly. The table also summarizes data gained with a mercury arc source, but in all examples significant amounts of the 2,4,5-T were decomposed.

C. The Laser Induced Decomposition of DNA.

Lasers can also be used to initiate the decomposition of DNA into small fragments. The samples were irradiated with 266 nm enegy from a quadrupled YAG, with the analytical methods being that of melting profile, low shear viscometry, and gel electrophloresis.

At the wavelength of irradiation, the chromophoricc group is one of the bases, all of which have a λ max. between 260 and 273 nm. The rate and degree of decomposition of the DNA is a function of the intensity and duration of irradiation. Figure 3 depicts the relative

Table I. Photolysis of 2,3,7,8-Tetrachlorodibenzo-p-dioxin (TCDD) in the Presence of 2,4,5-Trichlorophenoxy-acetic Acid (2,4,5-T)

Irrad Source/ λ (nm)	Power	Time, Hrs	2,4,5-T, %	TCDD, ppm
None			100	0.058
XeCl (308)	100mw	2	101	0.052
XeCl (308)	50mw	4	102	0.048
Nd/YAG (320)	50mw	4	100	0.037
N$_2$ (337)	50mw	4	96	0.057
Hg Arc (360)	400w	6	92	0.007**
Hg Arc (360)	400w	20.00*	76	-**
Hg Arc (360)	400w	7.25*	39	0.033***

*CH$_3$OH solvent   **Init = 0.033   ***Init = 0.390

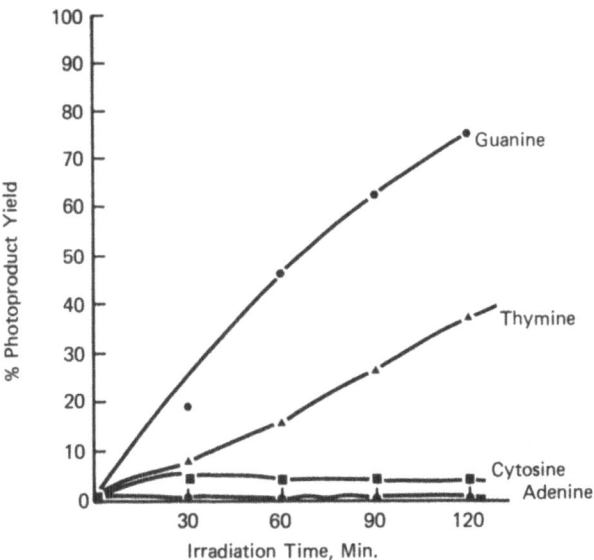

Figure 3.   Effect of irradiation time with 2.5mj pulse.

stability of the various bases at 2.5 mj/pulse. This data extrapo-
lates well with what would be expected in DNA when compared with a
mixture of bases. When these reactions are studied on various types
of DNA the analytical data indicates that smaller fragments of DNA
are actually produced as can be seen in a comparison shown in

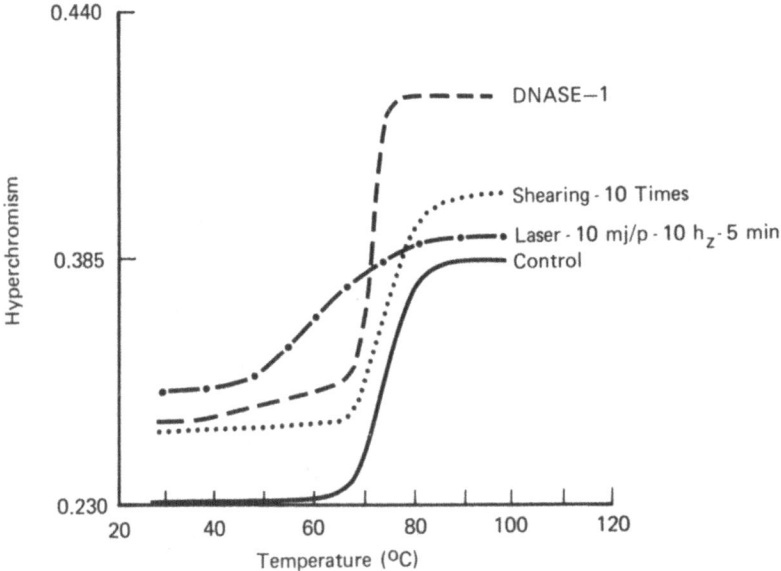

Figure 4.   Relative degree of DNA Cleavage.

Figure 4. This type of work has been previously reported on many systems.[4],[5],[6] The important note is that pulses with energy density $>10^{10}$ w/m^2 are necessary to affect cleavage rather than dimerization.[5]

This is an example of a selective decomposition which requires a bases but not due to wavelength but because of power density.

References

1. S. Toby & F. S. Toby, J. Phys. Chem. 84, 206-207 (1980).
2. S. Koda, Chem. Phys. Letters, 55, (2), 353-357 (1978).
3. E. M. Geiser, R. W. Johnson, USP. 4,287,038 (1981).
4. S. L. Shapiro, Ch 16, Biological Events Probed by Ultrafast Laser Spectroscopy.
5. T. N. Menshonkova, N. A. Simukova, E. I. Budowsky & L. B. Rubin, FEBS Let. 112 (2), 299 (1980).
6. G. G. Gurzadyan, D. N. Nikogosyan, P. G. Kryukov, V. S. Letokhov, T. S. Belogurov & G. B. Zavilgelskij, Photochemistry and Photobiology, 33 835 (1981).

PHOTOINITIATED CATALYSIS BY TRANSITION METAL CARBONYLS

D. J. Perettie, M. S. Paquette, and R. L. Yates
The Dow Chemical Company
Central Research
Midland, MI 48640

H. D. Gafney
Queens College
Flushing, NY 11367

Photocatalysis by metal carbonyl compounds has generated increased interest in view of the reaction systems which can be studied.[1] In the present work which is being pursued in our laboratory, catalysis by metal carbonyls is being studied based on photogeneration of thermal catalysts. Some of the systems under investigation are shown below.

PHOTOACTIVATED CATALYSIS

(1) $M \xrightarrow{h\nu} M^* \longrightarrow B$ Photogeneration of thermal catalyst B

(2) $A + B \longrightarrow C + B$

Hydrosilylation...

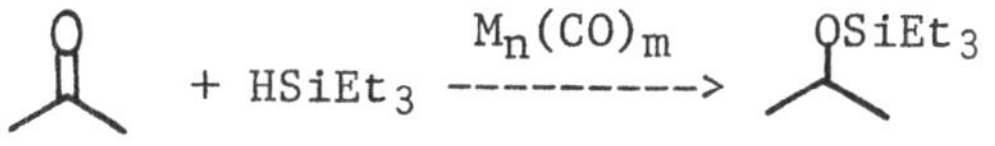

$M_n(CO)_m = Ru_3(CO)_{12}; Ir_4(CO)_{12}$
$Os_3(CO)_{12}; Re_2(CO)_{10}$

Olefin isomerization...

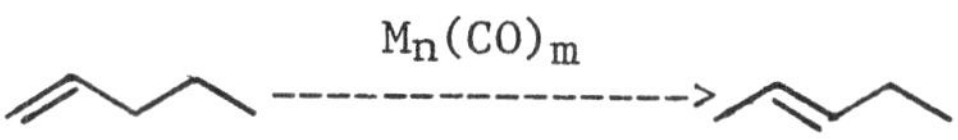

$M_n(CO)_m = Fe(CO)_5; Re_2(CO)_{12}$
$Os_3(CO)_{12}$

Ploymerization...

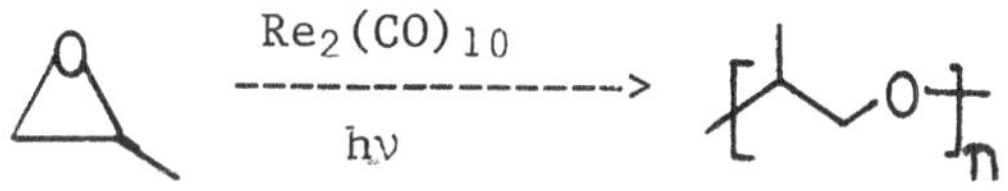

These systems are being investigated by standard techniques in addition to stop-flow FTIR and isolation employing Porous Vycor glass. The batch reaction system was employed both for the hydrosylation and epoxide ploymerization. In these cases, the reactants and catalyst were degassed and sealed in an ampule which was subsequently irradiated with either Mercury Arc lamps or various lasers. The products were analyzed by GC or LC with GPC being employed for the polymerization reactions.

The results of the work on the hydrosylation reaction have been described in detail elsewhere and therefore will not be elaborated on at this time.[2]

The photocatalyzed polymerization of alkylene oxides and in particular propylene oxide was also studied employing a batch technique. For this work, both mercury lamps and a XeF laser was employed to excite the metal carbonyl. The initial results for the polymerization of propylene oxide is shown in table 1. Since $Re_2(CO)_{10}$ appeared to have the greatest activity, similar to hydrosilations, a detailed study was initiated.

The initial experiments were performed at ambient temperature and the rate of conversion studied. The results of this work are shown in Figure 1. The conversion increased controllably until $\sim$ 40 min. at which times a rapid increase was observed which was accompanied by an exotherm. Upon analysis of the molecular weights for each experiment it was noted that as the degree of reaction

Table 1. Photocatalyzed polymerization of propylene oxide.

Catalyst	% Propylene Oxide Conversion[a]
$Re_2(CO)_{10}$	93
$Ru_3(CO)_{12}$	11
$Ir_4(CO)_{12}$	4
$Os_3(CO)_{12}$	8
$Fe_3(CO)_{12}$	< 1
$Mn_2(CO)_{10}$	0[a,b]
$Co_2(CO)_6(PPh_3)_2$	4
$Co_2(CO)_8$	5
$Rh_6(CO)_{16}$	20
$RhCl(CO)(PPh_3)_2$	0
$Mo(CO)_6$	0
$W(CO)_6$	< 1
$Cr(CO)_6$	0

[a] Irradiation time = 20 hrs.
[b] Irradiation time = 40 hrs.

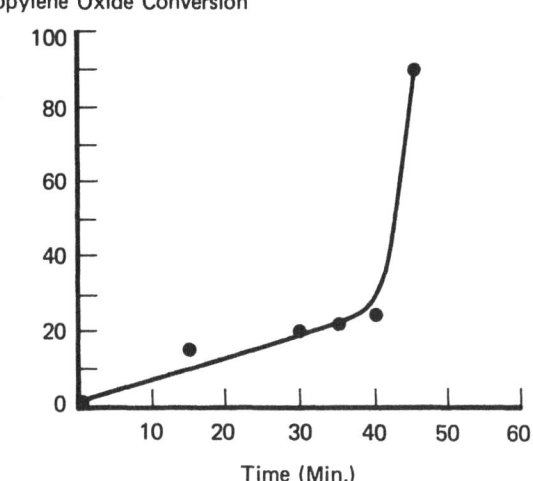

Figure 1.   The rate of propylene oxide conversion.

increased the molecular weight and distribution decreased which
indicates a photoinitiated depolymerization reaction.   This depolymerization reaction is intensity sensitive as can be seen in
Figure 2.   The general mechanism for both the polymerization and
depolymerization reactions is shown below.

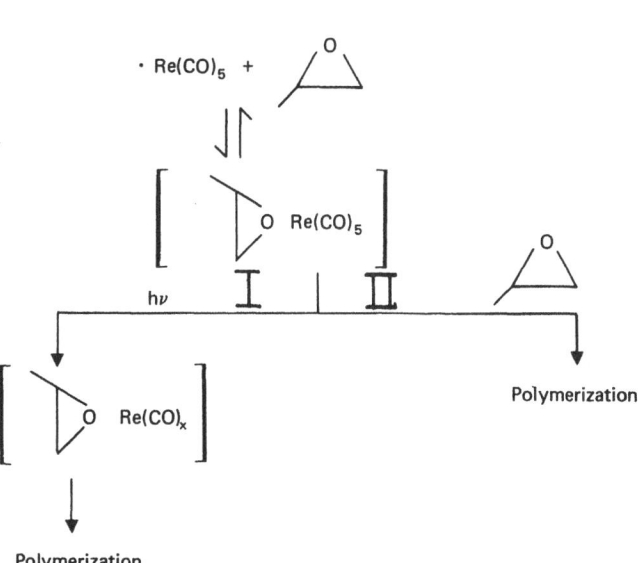

Figure 2.   The proposed mechanism for propulene oxide polymerization.

Path I is the general route leading to the lower m.w. species. This observation was substantiated by an increase in CO pressure on the reactor and both a slower and less exothermic reaction was experienced in addition to the resultant polymer possessing a higher molecular weight. Another experiment which was performed to substantiate the depolymerization pathway was to study elevated temperatures which also enhance the decarbonylation of the catalyst. As expected, the elevated temperatures resulted in enhancement of the lower molecular weight product.

Since this appeared to be a commercially interesting reaction, the question of supporting the metal carbonyls had to be addressed. Examples of this was to support the material on a silica. Both standard silica and a material referred to as Porous Vycor Glass (PVG) which is an unnealed, acid leached glass available from Corning Corp. were employed and resulted in good initial activity.

The studies on the PVG were expanded to include other metal carbonyl systems in which it is important to support the reactive species. This material has been studied extensively as an isolation technique to produce active catalyst which can be supported and utilized without resorting to ultra low temperatures.[4] An example of this phenomenon is depicted in Figure 3. The curves shown result from photolysis $W(CO)_6$ at 312 nm as a function of time. The band at 410 nm is indicative of a $W(CO)_5$ complex which is stable in the absence of light in vacuum for at least 48 hrs. This indicates that it is possible to ligate a reactive intermediate such as $W(CO)_5$. It is possible to form complexes of this species as shown in Figure 4,

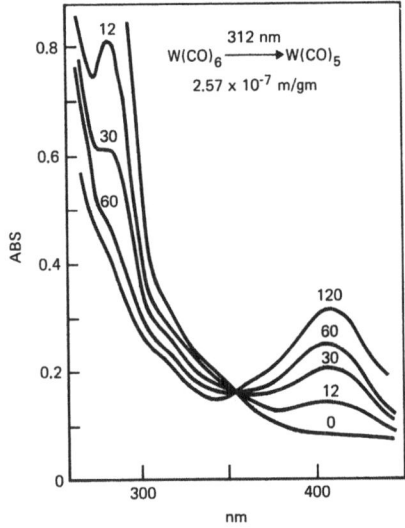

Figure 3. The spectrum of photolyzed $W(CO)_6$ on PVG.

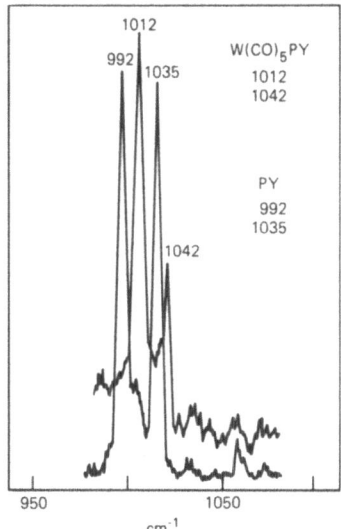

Figure 4.   The resonance raman spectrum of absorbed pyridene on PVG.

in this case with pyridine. A complex is evident since the pyridine bands (992 & 1035cm^{-1}) shift in the resonance raman spectrum to 1012 and 1042cm^{-1} when absorbed with the photolyzed $W(CO)_6$.

In addition to the complexation reactions described above, it is possible to study typical metal carbonyl reactions such as olefin isomerization. An example of this would be the isomerization of 1-butene on a photolyzed $W(CO)_6$ catalyst supported on PVG. Figure 5

Figure 5.   The isomerization of 1-butene with photo- lyzed $W(CO)_6$ on PVG.

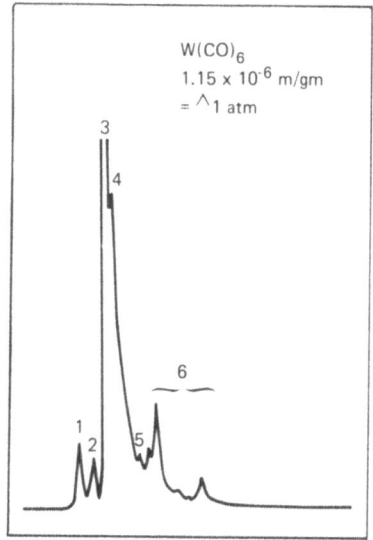

shows some preliminary results on the above isomerization reaction.
It is obvious from the data presented that it is possible to photo-
lyze a metal carbonyl to produce an "active" species which is stable
on PVG and ultimately study a catalytic reaction.

Another system which will be briefly covered is a technique for
studying metal carbonyl reactions in situ employing a rapid scanning
technique in a FTIR. The sample cell is positioned in the sample
chamber of the instrument with the necessary optics to direct either
a single or multiple laser pulses into the cell which contains both
the catalyst and the reactant in a solvent. An example of the assem-
bly is shown in Figure 6.

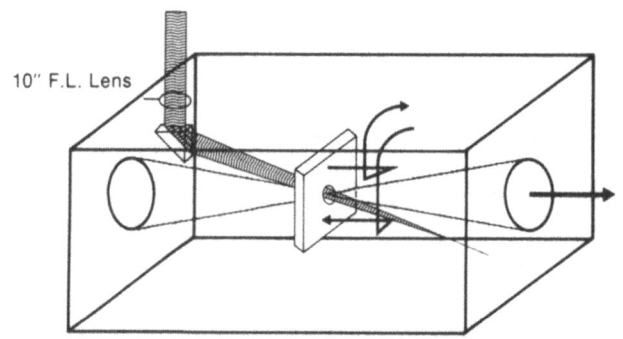

Rapid Scan FTIR Reaction Cell

Figure 6. Schematic of the liquid flow reactor.

Initial experiments on this system included the photolysis of
$Fe(CO)_5$ with 1-pentene in a pentane solution. The results of this
work are shown in Figure 7 where both single and multiple shot
(arrows) experiments are depicted. By monitoring the I.R. spectra
for a period of time it is possible to gain information into the
reactivity and turn-over number for many catalytic species.

The previous discussion covers many techniques to determine
the activity, reaction mechanism, and utility of many metal carbonyl
catalysts.

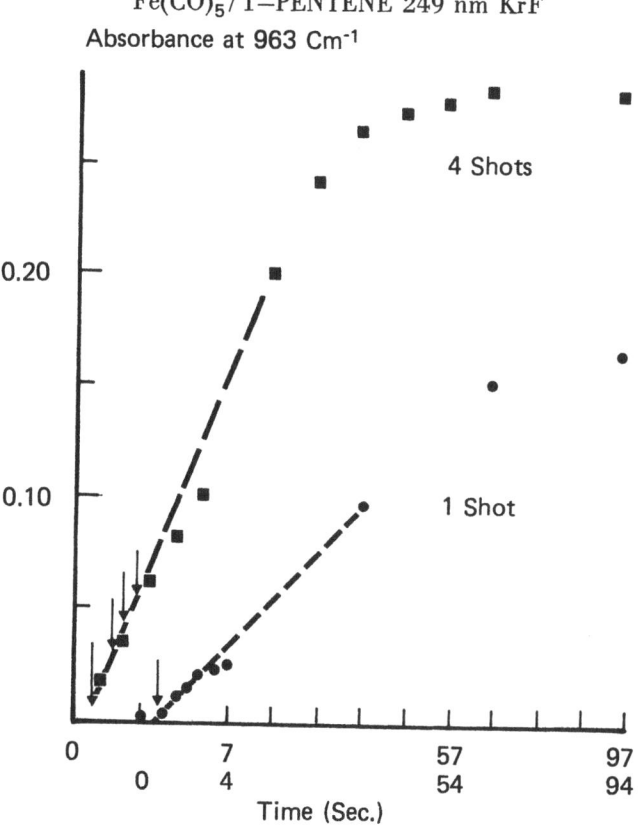

Figure 7.   The pulse dependent isomerization of 1-pentene with
            photolyzed $Fe(CO)_5$.

References:

1. G. L. Geoffroy and M. S. Wrighton, "Organometallic Photochemistry"
    Academic Press, 1979.

2. R. L. Yates, J. Cat., 78; 111-115 (1982).

3. M. S. Wrighton and D. S. Ginley, J. Am. Chem. Soc., 97, 2065
    (1975).

4. T. Kennelly and H. D. Gafney, J. Inorg. Nucl. Chem., 43, 2988
    (1981).

FORMATION OF C_2H_3Cl BY LASER-INDUCED RADICAL CHAIN REACTIONS

M. Schneider and J. Wolfrum

Max Planck-Institut für Strömungsforschung
D-3400 Göttingen, Böttingerstraße 4-8
and
Physikalisch-Chemisches Institut der
Ruprecht-Karls-Universität Heidelberg
D-6900 Heidelberg, Im Neuenheimer Feld 253

ABSTRACT

The laser induced radical chain elimination of hydrogen chloride from 1,2-dichlorethane (DCE) forming vinyl chloride (VC) has been studied in a flow reactor in the temperature range from 470 to 720 K. Product formation after the laser pulse was observed by a time-resolved UV-absorption technique. The rate constant of the decomposition of the 1,2-dichloroethyl radical

$$C_2H_3Cl_2 \xrightarrow{\quad M \quad} C_2H_3Cl + Cl$$

was measured to

$$k = (6.5 \pm 2) \times 10^{13} \times \exp(-83 \pm 3 \text{ kJ/mol/RT}) \text{ s}^{-1}$$

Effects of temperature, pressure and initial radical concentration on quantum yields and on k were determined.

INTRODUCTION

The special properties of rare-gas halide lasers such as defined energy, wave length and beam geometry, together with a short pulse length have opened numerous new ways for a specific electronic excitation of molecules[1]. This work describes the application of the RGH laser to investigate the mechanism and

259

kinetics of the photo-initiated chain reaction of 1,2-dichlor-
ethane (DCE) to form vinyl chloride (VC)[2]. The reaction is of considerable industrial importance, being the main route to production of vinyl chloride monomer feedstock for PVC manufacture.

EXPERIMENTS

As shown in Fig. 1 the experiments were made using a cylin-
drical fused quartz flow reactor with suprasil windows. The reactor was closely fitted in an aluminium tube enclosed in a tempera-
ture-controlled oven. A temperature gradient less than 1 % along
the reactor axis was observed. The photolysis source was a Lambda-
Physik EMG 501 laser operated at 193 nm (ArF), 222 nm (KrCl),
248 nm (KrF) and 306 nm (XeCl). Time resolved product analysis was
made by UV-absorption measurements with a HgXe-lamp, a 0.15 m monochromator (Oriel 7240) for preselection before the cell, a
second monochromator (0.3 m, Zeiss MW3) behind the cell and a
photomultiplier (EMI, 6256A) as detector. In addition gas samples
extracted from the reaction cell by gas syringe were analysed by
gas chromatography, mass spectrometry and in an infrared sample
cell.

The result of absorption measurements for DCE are plotted
semilogarithymically against wave length in Fig. 2. A steep rise
of the absorption cross section with shorter wave length and increasing temperature is observed.

Very small absorption cross sections were found above 250 nm.
Nevertheless substantial amounts of vinyl chloride can be formed
after irradiation with the KrF and XeCl laser lines. In order to
study this effect more quantitatively a series of experiments was
carried out varying the laser energy per pulse from 10 to 250 mJ
at fixed photolysis wave length, temperature and DCE pressure. In

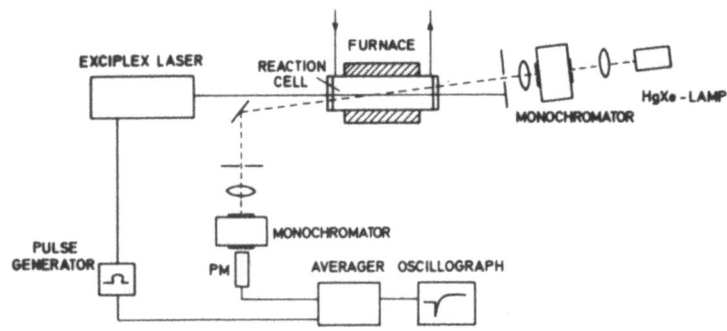

Fig. 1   Experimental arrangement for the investigation of laser
          induced radical chain reactions by product detection

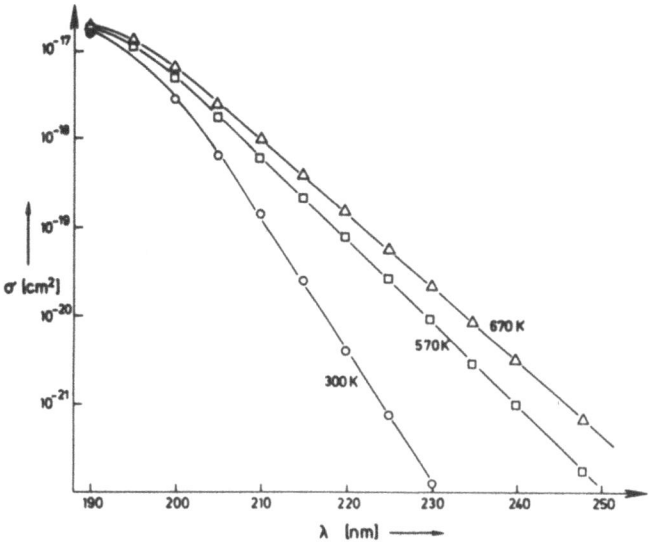

Fig. 2   UV-Absorption of 1,2 Dichlorethane at different
           temperatures

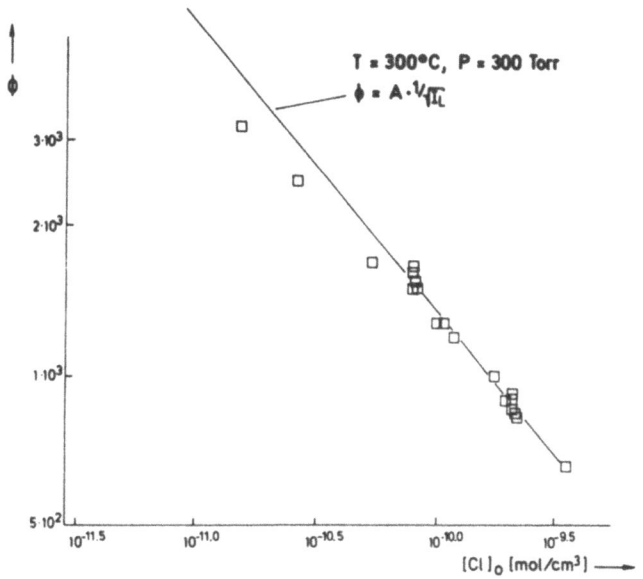

Fig. 3   Quantum yield of C_2H_3Cl formation as function of initial
           radical concentration

a double logarithmic plot of quantum yield versus initial radical concentration (see Fig. 3) one obtains a straight line with a slope of -0.5.

This negative square root dependence of the quantum yield can be explained by a Rice-Herzfeld mechanism[3]. The elementary steps involved are shown in Fig. 4. Good agreement is obtained between the measured and calculated rate of vinyl chloride formation.

An additional series of experiments was performed to study the temperature dependence of the quantum yield at a constant conversion rate per laser shot. This has the advantage of working with identical effects like adiabatic cooling or vinyl chloride inhibition and it could be achieved only by a drastic reduction of laser energy. At the highest temperature studied (720 K) and a low laser energy of 1 mJ/cm^2 we obtained quantum yields of more than 10^4. It was possible to extract the rate constant of the unimolecular reaction

$$C_2H_3Cl_2 + M \longrightarrow C_2H_3Cl + Cl + M \qquad (1)$$

from the time constant of the total reaction. A series of experiments with fixed temperature and pressure and varying irradiation intensity showed that the decrease of the chain length with higher initial radical concentration is paralleled by the increase of reaction time so that the arithmic product of both remains constant over the whole range of laser energy. Fig. 5 compiles data for the change at the unimolecular rate constant k_1 when an inert gas (N_2) was added to 60 torr of DCE up to a total pressure of 600 torr. The reaction rate is strongly pressure dependent in this region. In the double logarithmic plot log k/log p one gets a straight line with a slope of 0.52.

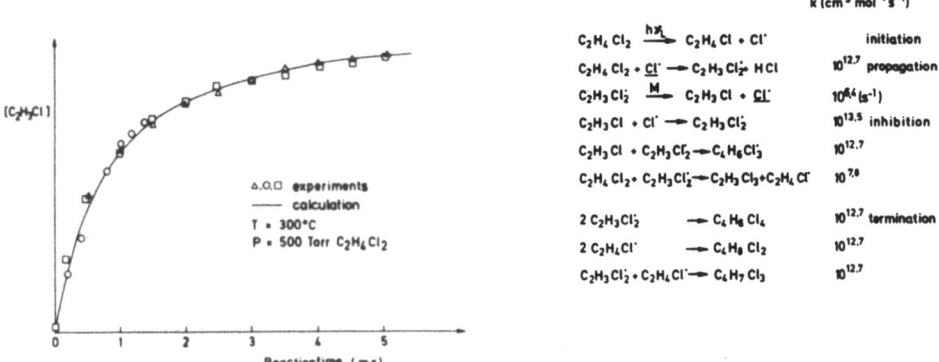

Fig. 4   Formation of C_2H_3Cl by UV-laser induced radical chain
          reaction

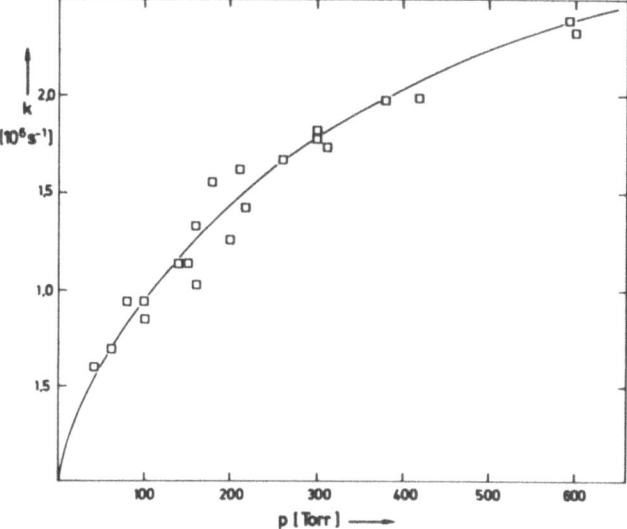

Fig. 5   Pressure dependence of VC formation by radical chain reactions

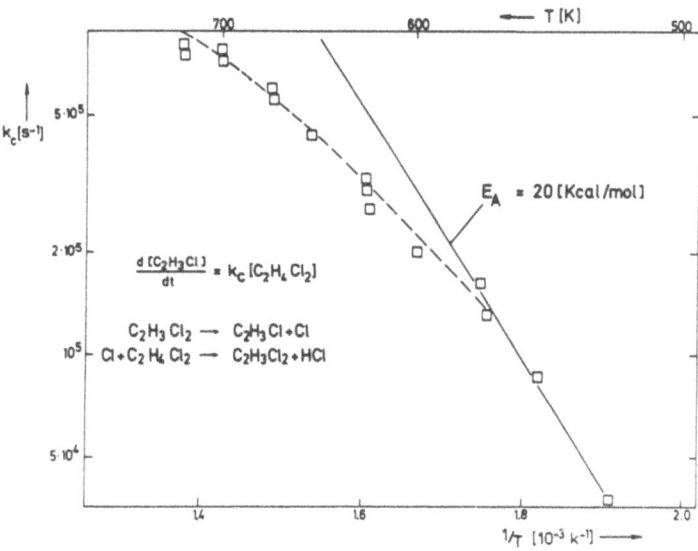

Fig. 6   Temperature dependence of VC formation rate by radical chain reaction

An Arrhenius diagram of the measurements at 300 torr DCE between 520 and 720 K (see Fig. 6) shows a decline of activation energy with temperature. This may be explained by two reasons. First, when moving to higher temperatures the unimolecular reactions shifts into its second order region where the activation energy is lower and secondly at higher temperatures k_1 is so accelerated that it is not anymore strictly the rate determining step though it is still a factor of 5 slower than the abstraction reaction.

$$Cl + C_2H_4Cl_2 \longrightarrow HCl + C_2H_3Cl_2 \tag{2}$$

Therefore it follows that the low temperature values should be taken to determine the Arrhenius parameters. In this region 520-570 K one gets the expression

$$k_1 = (0,5 \pm 2) \times 10^{13} \times \exp(-83 \pm 3 \text{ kJ/mol/RT}) \quad .$$

With the rate data on the pressure and temperature dependence of k_1 obtained in this investigation one can use the computer simulation model (see Fig. 4) to predict the effect of laser generated free radicals in the DCE to VC conversion at conditions used in the technical process (see Fig. 7). These calculations show that a significant increase in the conversion ratio can be obtained by laser photocatalysis.

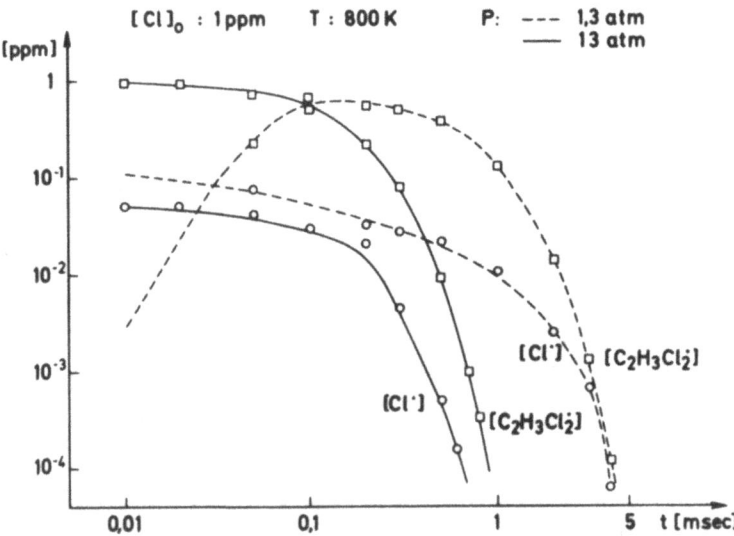

Fig. 7   Calculated time profiles for laser induced radical concentrations in the DCE to VC conversions at different reactor conditions

ACKNOWLEDGEMENT

The financial support of the Deutsche Forschungsgemeinschaft (Sonderforschungsbereich 93 "Laser Photochemistry") is gratefully acknowledged.

REFERENCES

1. Ch. K. Rhodes (Ed.), Excimer Lasers, Top. in Appl. Phys., Vol. 30, Springer Heidelberg 1979
2. J. Wolfrum, M. Kneba, P. Clough and M. Schneider, German Patent (DBP 2938 353 7)
3. F. O. Rice and K. F. Herzfeld, J. A. CS <u>56</u>, 284 (1934)

ADAM A.
Chemistry Dept.
University of Waterloo
Waterloo, Ont. N2L 3G1
(Canada)

ARECCHI F.T.
Istituto Nazionale d'Ottica
Largo E. Fermi, 6
50125 Firenze (Italy)

BAYRAKCEKEN F.
O.D.T.U.
Cevre Müh. Bölümü
Ankara (Turkey)

BACHMANN F.
Max-Planck-Institut
für Quantenoptik
Postfach 1513
8046 Garching (F.R.G.)

BERRY M.J.
Dept. of Chemistry and
Rice Quantum Institute
William Marsh University
P.O. Box 1892
Houston, Texas 77251
(U.S.A.)

BIANCIFIORI M.A.
ENEA - Ist. della Casaccia
Lab. Chim. Analitica (TIB/CHI)
S.P. Anguillarese Km 1,300
Roma (Italy)

BOSCHETTI A.
Dipartimento di Fisica
Università di Trento
38050 Povo (Trento)
(Italy)

BRISTOW N.J.
Inorganic Chemistry
University of Nottingham
Nottingham (U.K.)

CACCIATORE M.A.
Ist. Chim. Gen. ed Inorganica
Università
Via Amendola, 173
70126 Bari (Italy)

CARRINGTON A.
Dept. of Chemistry
The University
Southampton S09 5NH (U.K.)

CASTELLI F.
Istituto di Chimica Generale
Università
Roma (Italy)

CILIBERTO S.
Istituto Nazionale di Ottica
Largo E. Fermi, 6
50125 Firenze
(Italy)

DEGLI ESPOSTI C.
Ist. Chimico "G. Ciamician"

Via F. Selmi, 2
40126 Bologna
(Italy)

EMMI S.S.
Istituto FRAE-CNR
Via de' Castagnoli, 1
40126 Bologna (Italy)

FABIANI C.
ENEA - Ist. della Casaccia
Lab. Chim. Analitica (TIB/CHI)
S.P. Anguillarese Km 1,300
Roma (Italy)

FANTONI R.
ENEA - Lab. Spettroscopia Mol.
C.P. 65
00044 Frascati (Roma)
(Italy)

FEW G.A.
Group R & D Centre
British American Tobacco Co.
Regents Park Rd.
Millbrook, Southampton (U.K.)

FLAMIGNI L.
Istituto FRAE-CNR
Via de' Castagnoli, 1
40126 Bologna (Italy)

GIANINONI I.
CISE SpA
C.P. 12081
20134 Milano (Italy

GIL J.J.
Depart. de Optica
Facultad de Ciencias
Ciudad Universitaria
Zaragoza (Spain)

GOUGH T.E.
Dept. of Chemistry
University of Waterloo,
Waterloo, Ont. N2L 3G1 (Canada)

HAMILTON C.E.
Joint Institute for

Lab. Astrophys.
University of Colorado
Boulder, Colo. 80309
(U.S.A.)

Harrison R.G.
Dept. of Physics
Heriot-Watt University
Riccarton,
Edinburgh EH14 4AS
(U.K.)

HENKE W.E.
Inst. f. Physikalische u.
Theoretische Chemie, TUM
Lichtenbergstr. 4
8046 Garching (F.R.G.)

HESS P.
Physikalisch-Chemisches Inst.
der Universität
Neuenheimer Feld 253
6900 Heidelberg (F.R.G.)

HUMPHRIES M.R.
Chemistry Dept.
Heriot-Watt University
Riccarton, Edinburgh EH14 4AS (U.K.)

HUTSON J.M.
Dept. of Chemistry
University of Waterloo
Waterloo, Ont. N2L 3G1 (Canada)

JACOBS A.
MPI f. Strömungsforschung
Böttingerstr. 4/8
3400 Göttingen
(F.R.G.)

KALDOR A.
Exxon Research and Engineering Co.
Linden, N.J. 07036
(U.S.A.)

KENNEDY R.
Dept. of Chemistry
The University
Southampton S09 5NH (U.K.)

KOMPA K.L.
Max-Planck-Institut für
Quantenoptik
Postfach 1513
8046 Garching (F.R.G.)

Krieger W.
Max-Planck-Institut für
Quantenoptik
Postfach 1513
8046 Garching (F.R.G.)

LEACH S.
Lab. Photophysique Moléculair
Bâtiment 213 - Univ. de Paris Sud
91405 Orsay Cédex, (France)

LEWIN A.K.
Dept. of Chemsitry
University of Waterloo
Waterloo, Ont. N2L 3G1
(Canada)

MAZUR E.
Harvard University
Division of Applied
Sciences, Pierce Hall
Cambridge, Ma. 02138,
(U.S.A.)

ter MEULEN J.
Fysisch Laboratorium
Katholieke Universiteit
Toernooiveld
Nijmegen (The Netherlands)

MORALES P.
ENEA-Ist. Casaccia
Laboratorio Tecnologie
Speciali, COMB/MEPIS
C.P. 2400
00100 Roma (Italy)

MORELLI G.
Ist. Chimica delle Macro-
molecole,   C.N.R.
Via Bassini, 15/A
20133 Milano (Italy)

PERETTIE D.J.
The Dow Chemical Co.
M.E. Pruitt Research Center
Midland, Michingan 48640
(U.S.A.)

PIMENTEL G.C.
Lab. of Chemical Biodynamics
University of California
Berkeley, Ca. 94720 (U.S.A.)

PROUT J.M.
The Physical Chemistry Lab.
Oxford University
South Parks Rd.
Oxford OX1 3QZ (U.K.)

RAM S.
Mathematics Dept.
St. Andrews University
St. Andrews KY16 9SS
(U.K.)

REBENTROST F.
Max-Planck-Institut für
Quantenoptik
Postfach 1513
8046 Garching
(F.R.G.)

RIDI N.
Istituto nazionale di Ottica
Largo E. Fermi, 6
50125 Firenze
(Italy)

RUBIO ALVAREZ M.A.
U.N.E.D.- Facultad de Ciencias
Apartado de Correos 50 487
Madrid (Spain)

SCHMATJKO K.J.
Kraftwerk Union AG
Dept. B 312
Hammerbacher Str. 12-14
8520 Erlangen (F.R.G.)

SCHRÖDER H.
Max-Planck-Institut für
Quantenoptik
Postfach 1513
8046 Garching (F.R.G.)

SCOLES G.
Dept. of Chemistry
University of Waterloo
Waterloo, Ont. N2L 3G1
(Canada)

SETHI D.S.
Chemistry Dept.
University of Bridgeport
Bridgeport, Ct. 06601
(U.S.A.)

SVELTO O.
Centro Elettronica
Quantistica del C.N.R.
Piazza Leonardo da Vinci, 32
20133 Milano (Italy)

TERRANOVA M.L.
Ist. Chim. Gen. ed Inorganica
Piazzale Aldo Moro, 5
00185 Roma (Italy)

ULIVI L.
Istituto Nazionale di Ottica
Largo E. Fermi, 6
50125 Firenze (Italy)

WANNER J.
Max-Planck-Institut
für Quantenoptik

Postfach 1513
8046 Garching (F.R.G.)

WELGE K.H.
Fakultät für Physik
der Universität Bielefeld
Postfach 8640
4800 Bielefeld 1
(F.R.G.)

WOLEJKO L.
Instytut Fizyki
Ul. Grunwaldzka 6
60-780 Poznan (Poland)

WOLFRUM J.
Max-Planck-Institut für
Strömungsforschung
Böttingerstr. 4/8
3400 Göttingen (F.R.G.)

WRIGHT J.C.
Dept. of Chemistry
University of Wisconsin
Madison, Wi 53706, (U.S.A.)

YUSTE M.
U.N.E.D. - Facultad de Ciencias
Apartado de Correas 50 487
Madrid (Spain)

ZEN M.
Dipartimento di Fisica
Universitá di Trento
38050 Povo, Trento
(Italy)

Absorption measurements, 59
  in the atmosphere, 90
Addition of fluorine to
        olefins, 188
Analytical chemistry, 1, 57
  present methods, 57
Angular distribution, 195, 219
Angular momentum polariza-
        tion, 196
Atom-molecule reaction, 194,
        199
Autoionization, 164

Bistability, 178
BOXCARS, 83

Catalysis, 211, 251
Chaotic dynamics, 171
Chemical analysis,
  see analytical chemistry
Chlorination of 1,1 -
        Dichloroethane, 240
Chlorophyll-Quinone photo-
        reaction, 225
Coherent anti-Stokes Raman
        spectroscopy (CARS),
        82
Coherent nonlinear methods,
        81
Coriolis resonances in CH_3F,
        138

Decomposition reactions,
        245, 247
Dehydrochlorination of
        1,2-Dichloroethane,
        242

Differential absorption, 91
Difference-frequency laser,
        117
DNA decomposition, 247
Doppler spectroscopy of H
        and D atoms, 107

Electron storage ring, 38
  synchrotron, 38
Energy transfer in poly-
        atomic molecules,
        134, 146
Excimer lasers, 20
Excited state processes in
        molecules, 36

Fluorescence scattering, 90
  spectroscopy, 67
Free electron laser, 50
  possible uses, 53
  properties, 52
Free radical reactions, 239

Infrared molecular gas
        lasers, 25
Infrared laser magnetic re-
        sonance, 118
Intracavity absorption spec-
        troscopy, 63
Internal population distri-
        bution, see inter-
        nal product state
        analysis
Internal product state analy-
        sis, 126, 194, 201,
        204, 219

Internal (continued)
  Λ Component specific distri-
      bution, 130, 205
Inverse Raman scattering, 85
Ionization of atoms, 25, 96
  of molecules, 161, 24
Ion fragmentation, 163
Isomerization of 1-butene,
      255

Laser chemistry, 1
  systematics of, 9
Laser diagnostics of chemical
      kinetics, 3, 193, 199
Laser-induced fluorescence,
      67, 193,
  in atoms, 67, 95
  in molecules, 68, 124,
      135, 195, 200, 217
Laser-initiated reactions, 6
  addition reactions, 187
  atom-molecule reactions,
      199
  chlorination, 240
  decomposition reac-
      tions, 245, 247
  dehydrochlorination, 242
  free radical reactions, 239
  isomerization, 255
  multiphoton excited
      reactions, 183
  photocatalysis, 251
  photolysis, 225
  polymerization, 252
  radical chain reactions, 259
Laser sources, 15
  frequency tunability, 18
  output, 18
  survey, 16
  wavelengths, 17
LIDAR, 90

Matrix isolation, 70, 187
Mie scattering, 90
Mode selective energy trans-
      fer in CH_3F, 133
Mode specific excitation of
      fluorine-olefine
      reactions, 187

Molecular gas lasers, 18
  optically pumped, 25
Molecular ions, 117
Molecular ion beams, 119
Multiphoton absorption (MPA),
      152
Multiphoton dissociation
      (MPD), 6, 141, 151
Multiphoton excitation
      (MPE), 6, 151, 152, 171
  of reactions of F_2CCHF, 183
Multiphoton fragmentation, 164
  statistical models, 166
Multiphoton ionization,
      161, 211
  of atoms, 75
  of molecules, 76, 211
Multiplex CARS, 84

1/f Noise, 178
Nonlinear spectroscopic tech-
      niques, 81

Optically pumped molecular
      gas lasers, 25
  specific systems, 26
  experimental considera-
      tions, 29
  theoretical considerations, 30
Optoacoustic Raman spectro-
      scopy (OARS), 84
Optoacoustic spectroscopy, 59
Ozone measurement in the
      atmosphere, 93

Photoacoustic Raman spectro-
      scopy (PARS), 84
Photocatalysis, 251
Photodissociation, 163
  of molecular ions, 120
  of NO_2, 123
2-photon ionization, 104, 109, 163
2-photon spectroscopy of H_2, 111
3-photon ionization, 163
Photosynthesis, 225
Polymerisation of propylene
      oxide, 252

Radical chain reaction, 259

Raleigh scattering, 90
Raman gain spectroscopy, 84
  loss spectroscopy, 84
  scattering, 90
Resonance CARS, 83

Shpol'skii effect, 70
Single atom detection, 75, 95
Stimulated Raman scattering,
    85
State specific chemistry, 2,
    193
State to state kinetics,
    2, 193, 196
  techniques of, 194
Supersonic jet, 71, 218
Surface processes, 207, 215
  laser diagnostics of, 217
  laser stimulated desorp-
    tion, 208
  laser stimulated reaction,
    210
  laser heating, 211
  laser induced ionization,
    211
  catalysis, 211
  scattering, 216
    elastic, 216, 223
    inelastic, 216, 223
    trapping/ desorp-
    tion, 216, 223
    reactive, 216
Synchrotron radiation, 35

comparison with VUV lasers,
    41, 104
characteristics of, 42
sources of, 37

Thermal lensing effect, 61
Thermal lensing spectroscopy,
    61
Translational reagent
    excitation, 199

Undulators, 48
  spectral characteristics,
    49
UV laser-induced excita-
    tation, 7, 161, 199
UV laser-induced photoioni-
    zation, 7, 161
UV laser-induced fragmentation,
    7, 161

Vibrational photochemistry, 6
Vinyl chloride synthesis, 259
VUV laser, 40, 103
  comparison with synchro-
    tron radiation, 41,
    104
VUV sources, 40
  characteristics of, 42
VUV spectroscopy of atomic
    hydrogen, 104
  of molecular hydrogen, 109
V-V process, 137, 146

Wigglers, 47